FUTURE TIME PERSPECTIVE AND MOTIVATION

Theory and Research Method

Louvain Psychology Series: Studia Psychologica

Series founded in 1953 by A. Michotte and J. Nuttin
Editorial board: L. Delbeke and J.M. Nuttin, Jr.

The Louvain Psychology Series "Studia Psychologica" is the continuation of the "Etudes de Psychologie" founded in 1912 by Albert Michotte.

FUTURE TIME PERSPECTIVE AND MOTIVATION

Theory and Research Method

JOSEPH NUTTIN
University of Louvain

With the collaboration of Willy Lens

Published jointly by

LEUVEN UNIVERSITY PRESS
LAWRENCE ERLBAUM ASSOCIATES

Revised edition in English of:

Motivation et perspectives d'avenir, published by Presses Universitaires de Louvain, Louvain 1980.

Copyright © 1985 by Leuven University Press and Lawrence Erlbaum Associates, Inc.
All rights reserved. No part of this book may be reproduced in any form, by photostat, microform, retrieval system, or any other means, without the prior written permission of the publisher.

Leuven University Press v.z.w.
Krakenstraat 3
B-3000 Leuven/Louvain (Belgium)
ISBN 90 6186 172 1
D/1984/1869/10

Lawrence Erlbaum Associates, Inc., Publishers
365 Broadway
Hillsdale, New Jersey 07642 (U.S.A.)
ISBN 0-89859-611-4

CIP KONINKLIJKE BIBLIOTHEEK ALBERT I

Nuttin Jozef, 1909—
Future time Perspective and Motivation: Theory and Research Method/Joseph Nuttin; with the collaboration of Willy Lens. — Leuven: Leuven University Press; Hillsdale, New Jersey: Lawrence Erlbaum Associates, 1985. — 231 p.; 24 cm.
Louvain Psychology Series: Studia Psychologica.
Revised edition of: Motivation et Perspectives d'Avenir.
Oorspr. uitg.: Louvain: Presses Universitaires de Louvain, 1980.
ISBN 90 6186 172 1
SISO 100   UDC 159-94
Onderwerpen: Tijd; Motivatie (psychologie)
Printed in Belgium

CONTENTS

Preface 9

Chapter 1. A theory of time perspective 11

1. The temporal dimension of behavior 12
2. The origin of the future in behavior 13
3. The construction of time perspective 15
   - Perspective 16
   - Temporal signs and their origin 17
   - The presence of objects in time perspective 19
4. An operational definition: Measuring time perspective 23
5. Additional aspects of time perspective 26
   - An active time perspective 26
   - Realism of time perspective: Temporal integration and time competence 28
   - Experiential learning and time perspective 31
6. The impact of future time perspective (F.T.P.) 32
   - The content of F.T.P.: Its impact on behavior 33
   - The impact of the temporal extension of F.T.P. 36

Conclusion 38
Notes 40

Chapter 2. The motivational induction method (MIM) 43

1. The instrument 45
2. Motivational Content Analysis 48
   - The purpose of motivational content analysis 49
   - The main categories of motivational objects and interactions 50
3. Validity of the MIM 54
   - Face validity of the MIM 54
   - Validity studies 57
4. MIM reliability and stability of motivation 59
   - Coding reliability 60
   - Stability of MIM-data 61

6 Contents

5. Additional possibilities of data-collecting by means of the MIM	63
- Frequency of expression and subjective intensity of motivation	64
Notes	65

Chapter 3. Measuring time perspective: the temporal code — 67

1. The basic principle	67
2. Justification of the "average" localization	68
3. The life periods of subjects	70
4. The temporal code: The time of objects	73
- Temporal scale in terms of calendar units	75
- Temporal scale in terms of social and biological units	75
- The open-present	75
- References to the past	76
- Atemporal objects	76
5. Reliability of the temporal code and stability of results	77
Conclusion	77
Note	78

Chapter 4. Measuring the extension of the F.T.P. — 79

1. Future time perspective profile	79
2. Future time perspective index	81
3. The mean future time perspective	81
A. The median rank of F.T.P.	82
B. The mean extension of F.T.P.	82
- Calendar periods	84
- Social-clock periods	84
Notes	89

Chapter 5. Attitudes towards the personal past, present, and future — 91

1. The time attitude scale (T.A.S.)	91
2. The revised time attitude scale: A multi-dimensional scale for the attitude towards the future	97

Contents 7

Chapter 6. **Manual of time perspective analysis**
(by J.R. Nuttin & W. Lens) 101

1. Coding units and their context 104
2. Description of the code 107
3. The coding technique 111
4. List of examples 124
   (Detailed Contents: see p. 101)

Chapter 7. **Manual of motivational content analysis**
(by J.R. Nuttin & W. Lens) 133

1. Principles of our content analysis 135
2. The structure of the code 137
3. The coding technique 139
4. List of examples 167
   (Detailed Contents: see p. 133)

Appendix A. **MIM instructions and list of inducers** 177

1. Suggestions to the instructor 177
2. Instructions to the subjects 178
3. Complete list of MIM inducers 180
4. Shorter form A of MIM inducers 181
5. Shorter form B of MIM inducers 182

Appendix B. **Inventory of motivational categories and Inventory of motivational objects** 185

Appendix C. **Computer analysis of MIM data**
(by W. Lens & A. Gailly) 201

References 209

Publications by members of the Research Center for Motivation and Time Perspective 219

Author Index 227

Subject Index 231

PREFACE

The present volume is a theoretical and methodological supplement to my book *Motivation, planning, and action: A relational theory of behavior dynamics* (Nuttin, 1984). In that book, I tried to show among other things that behavior dynamics in man, due to their interaction with higher cognitive functioning, are processed into goals and means-end structures, i.e. behavioral projects or plans. The objects of these plans or goals are temporally localized in a more or less distant future, thus creating a future time dimension - a time perspective - in the subject's behavioral world. The relationship between future time perspective and cognitively processed motivation is the theoretical underpinning of our method for studying and measuring future time perspective.

The main purpose of the theoretical section of this book (Chapter I) is to contribute to the integration of the future time dimension in the study of human motivation and behavior. Although ignored in many psychological textbooks, the future is an essential component of a person's behavior and his behavioral world. The ability to construct far-distant personal goals and to work toward their realization is an important characteristic of human beings. It is implied in the achievement of major projects where long-term instrumental steps are required and where the regulating impact of a goal is necessary from the very beginning of the enterprise. It seems plausible to admit that the psychological inability of some people to achieve long-term projects is related to a lack of future time perspective. One thinks of people for whom in many countries the necessity of immediate satisfaction of physiological needs continuously dominates behavior. Psychological treatment gradually extending their future time perspective, paralleling the development of their economic conditions, may be a necessary step in improving their motivation and achievement. On the other hand, several groups, such as juvenile delinquents and elderly people, seem to have problems related to future time perspective. Giving people a future to live in and to work for may become a crucial challenge for applied psychology in the society of tomorrow. Similar problems exist in youngsters who, lacking a future time perspective, are unable to perceive the instrumental link between their present studies and a far-distant career, so that motivation remains low. The impact of future time perspective on working and economic behavior in general is no less important, whereas even in the area of human relations, with regard to parents as well as to social leaders, a lasting influence on people implies the ability to evaluate the long-term outcomes and consequences of one's intervention.

It thus appears that the study of future time perspective promises to be important in the field of applied as well as of theoretical psychology. In contrast to physical time, the psychological future is constructed at the representational level of cognitive functioning. Therefore, it should be emphasized that the *level of reality* in a person's future time perspective is a basic requirement for its positive impact on current behavior. In fact, this reality level will determine whether the future fulfills its regulating function in purposive behavior or is mainly an attempt to escape into sterile fantasy.

The main part of this book is devoted to a two-step method for studying the content and density of a person's future time perspective and, most importantly, for measuring its depth or extension. A technique for getting samples of subjects' motivational goal objects and for localizing these objects in their respective future time periods has been developed (Chapters 2 - 4). In Chapter 5, a multidimensional scale for measuring the subjects' *attitude* towards the past, present, and future is described as an auxiliary approach to the study of some affective and other aspects of psychological time. Finally, two *Manuals* for the application of the two-step method are to be found in Chapters 6 and 7, and some auxiliary techniques are described in *Appendices A* and *B*.

This English publication is a revised edition of a book published in French (Nuttin, 1980). Thinking back about the many young psychologists who collaborated in our *Research Center for Motivation and Time Perspective*, I wish to express to all of them my warm appreciation for their contribution to the methods worked out in this volume. A list of publications by some of them is to be found at the end of our *List of references*. In particular, I am grateful for the collaboration of Dr. Thérèse Noterdaeme whose contributions in the early stages of our work are highly appreciated and to Dr. Willy Lens who is co-author of Chapters 6 and 7. Our secretary, Miss O. Wieërs, was helpful in many ways in the production of the book.

Louvain, January 1984.   J.N.

CHAPTER I

A THEORY OF TIME PERSPECTIVE

The concept of *time perspective* is a rather ambiguous one, and its integration in behavioral sciences is still very limited. Therefore, we will first try to specify its theoretical status and identify the processes involved. This will show us how the intentional objects of cognitive and motivational processes define the content of the temporal perspective, while their inherent 'temporal signs' constitute its time dimension.

A clear distinction has to be made between three different aspects of psychological time which often are referred to by the same term *'time perspective'*. The first is *time perspective* proper, which is essentially characterized by its extension, density, degree of structuration, and level of realism as will be shown in the following sections. The second is *time attitude* and refers to a subject's more or less positive or negative attitude towards the past, the present, and the future. Thus, one may anticipate one's future as offering more opportunities than the past (optimistic future attitude) and have a negative attitude towards the present. *Time orientation*, finally, refers to the preferential direction in a subject's behavior and thought insofar as it is predominantly oriented towards objects and events in the past, the present, or the future. Thus, one may assume for instance that most young people are future-oriented, whereas older people are past-oriented. In this book, our main concern is with time *perspective* in its proper and limited sense. The concept of time *attitude* will be discussed in Chapter V. As to *time orientation*, we refer to Nuttin et al. (1979), De Volder (1979), Cottle (1976) and Hoornaert (1973).

I. THE TEMPORAL DIMENSION OF BEHAVIOR

A behavioral act necessarily happens in the present: it is always *now* that I write, perceive, speak, think, walk, etc. It is even because of the coincidence of an event with the actual perception of a subject that an event is said to happen in the psychological *present*. Objectively speaking, two events may be *simultaneous*, but the psychological *presence* refers to the coincidence between an event and an actual behavioral operation. This present moment is as fleeting and continuously transitory as the subject's perceptual act itself (compare Fraisse, 1981).

Because each behavioral action happens in the present, only those variables that are active at that moment affect and help to explain that action. That is what Lewin (1935) meant with his *principle of contemporaneity* and his concept of the ahistorical determination of behavior. Nevertheless, psychologists studying psychological time (as Lewin did), stress the fact that not only the present state of affairs, but also future and past events in the frame of a subject's time perspective co-determine behavior. Our purpose is to show how.

Two considerations are important in that respect. First, time perspective is constituted by objects or events that exist on the *representational* or *cognitive* level of behavioral functioning. Second, the *objects* of an individual's cognitive representation are not tied to the present moment at which the representational act takes place. Thus, the conference I am thinking about at the present moment may be a conference that took place a year ago; it may be also the one I am planning to attend next year. The object itself is an essential element of the representational act: thinking, planning, and remembering are impossible without an object that constitutes the intentional content of those cognitive acts. Therefore, the object of a representational act is actually present in the subject's psychological activity, although the temporal localization of that object - the conference - may be either in the future or in the past. In the frame of time perspective, future and past events have an impact on present behavior to the extent that they are actually present on the cognitive level of behavioral functioning. This means that acknowledging the effect of time perspective in behavior depends on the recognition of the role of cognitive processes in behavioral functioning (1).

It is obvious that the impact of time perspective on present behavior is different from the learning effect. The fact that I speak and write at this moment the way I do, is the result of previously performed actions. Due to learning processes, the past is present in what we do at each moment. The study of the temporal dimension of behavior has mostly been limited to this unilateral learning effect. In it, however, the past does not exist any longer *as past*. The same holds for semantic memory (Tulving, 1972). The individual knows, for example, the meaning of a word, without consciously remembering the past acts on which that knowledge is based. We conclude: Only past and future events that form the content of actual cognitive functioning are psychologically present as belonging to the past or to the future; as such they constitute an individual's time perspective and may influence present behavior.

II. THE ORIGIN OF THE FUTURE IN BEHAVIOR

Contrary to the effect of the past, the role of the future is not generally accepted in behavioral science. Scientific methodology cannot accept that not yet existing (future) events are already active. A psychology that does not take into account the fact that, on the cognitive level, the future may be part of the present can not consider future time perspective either. That is why for some psychologists the concept of future time perspective is of no use in explaining behavior. On the other hand, those who do acknowledge the future time perspective as a behavioral determinant often consider it as an effect of the past.

Indeed, the future has been introduced in general psychology, particularly since Tolman (1932), through the notion of anticipation or expectancy. Expectancy or anticipation is conceived as based on a succession of events that happened in the past. Ringing the bell has been followed, in the past, by the presentation of food. Therefore, it is said, a subject hearing the bell later on will *anticipate* food. Thus, the future is conceived in terms of memory and conditioning; it is based on the past. This opinion is summarized by Fraisse (1967): "Due to memory, we are able to reconstruct the succession of previously experienced changes, and to anticipate such changes in the future."

In our opinion, however, this thesis does not explain expectation and future-orientedness in general. Reconstructing a succession of events that were perceived or experienced in the past is nothing more than recalling that, *in the past*, event B followed event A. But such memory is not yet anticipation. The change in temporal orientation from the past (memory) to the future (anticipation) cannot be explained in terms of memory or conditioned effects of the past. From his earliest experiments Pavlov observed that conditioning requires that the animal directs its attention to what will be happening. Such a future-oriented attention results from a state of alertness or activation caused by a need state. The need involved can just as well be curiosity as hunger. Due to this motivational state and the attention resulting from it, the psychological organism will detach itself in some sense from the present and become oriented towards something that is not yet present. The orientation reflex (attention) directs the subject not only to the already experienced change in stimulation (the past), but also - as always in a state of alertness - towards what may happen afterwards (the future). On the contrary, in a state of need satisfaction or pleasure, the individual is rather locked up in the present, as many authors have shown. For that reason, the unconditioned stimulus (food) should not precede the conditioned stimulus (the bell), but must *follow* it. When the food comes first (need satisfaction), the animal is no longer oriented towards 'what is not yet present' or towards 'what follows'.

We do not claim, however, that the past is without any effect on the orientation towards the future. The fact that event B - and not any other event - is anticipated when event A is perceived, results from past experiences. In this specific meaning, the *content* of the future can be seen as a projection of the past; but orienting oneself towards the future is a new and original phenomenon that originates in a state of motivation or need. At first, it is only a very general orientation (alertness), which is limited to the immediate future, as is the case with the orienting reflex in animals. Due to the higher cognitive functions and the cognitive processing of human needs into goal objects and means-end structures (Nuttin, 1984), anticipation becomes more and more detached from the present situation and deep perspectives develop. The construction of such human time perspectives is what we have to discuss next.

III. THE CONSTRUCTION OF TIME PERSPECTIVE

It is not our intention to write a historical review of the concept of time perspective and research related to it (2). We only would note that Lewin (1931) in a publication on the 'level of realism' referred not only to the 'spatial extension', but also to the 'temporal extension' of the psychological present life space (*psychologische Umwelt* or *psychologischer Lebensraum*). Not yet using the term 'time perspective', he discussed how the child's narrow horizon of the present (*der enge Horizont des Gegenwärtiges*) progressively extends in spatial (social) and temporal (mostly future) directions.

Lewin does not conceive of this growing extension as a pure and simple effect of intellectual development, but as the expression of the child's personal and constructive activity (*Selbsttätigkeit* and *Produktivität*). After the publication of Frank's (1939) cultural-philosophical article on 'time perspectives', Lewin (1942; 1946) adopted the term. He defined it as "the totality of the individual's views of his psychological future and his psychological past existing at a given time" (Lewin, 1952, p. 75) (3).

Since that time, especially after 1950, time perspective became the topic of much quantitative differential research (e.g. future time perspective with relation to social class, age, intelligence, achievement motive, delinquency, psychopathology, etc.). But the avalanche of studies created a great terminological confusion. In one of the first studies, Leshan (1952) uses the term *temporal goal orientation* in the sense of what we would now call the extension of future time perspective. More recently, Trommsdorff et al. (1979) still conceive of *distance* and *extension* as aspects of future *orientation*, adding the term 'time perspective' between parentheses; in the same article, some aspects of time attitude are also attributed to time orientation. As a result, scores of very heterogeneous studies on various aspects of psychological time, and using very different measurement instruments, are grouped under the heading *time perspective*, while investigations of time perspective proper are called time orientation studies. Therefore, the comparability of their data is highly questionable.

One of the reasons why the concept of *time perspective* became so ambiguous is that its parallelism with the notion of *space perspective* and its extension has been neglected. Some authors (e.g. Wohlford, 1964) have suggested abandoning the term because of its ambiguity and using the term 'personal time', which is an even more general concept. We would suggest using the term *time perspective* in its strict sense and distinguishing it clearly from the two related concepts of *time orientation* and *time attitude*, as noted above. Let us, therefore, consider the origin of the concept of *perspective* in general and describe the psychological construction of time perspective.

a. Perspective (4)

Many concepts about time have their origin in our thinking about the spatial dimension. This is the case also for the concept of *time perspective*. Space perspective refers essentially to the impression of depth that is created by a certain arrangement of graphic representations on a surface. The invention or reinvention of linear perspective by Brunelleschi and Alberti in the 15th century in Florence has been studied by Edgerton (1975). The confusion within the concept of time perspective results from disregarding the *impression of depth* as a fundamental aspect of perspective. In the visual perception of the real world, the perceived depth corresponds to objective distances that can be directly experienced by a subject *moving* frome one object to the next. In the temporal domain, these 'distances' are temporal intervals that can be experienced directly in the 'lived-through succession' of events. Analogous to space perspective, time perspective consists of the 'mental perception', at a certain moment in time (the psychological presence corresponding to the temporal here), of events that, in reality, happen in temporal succession and with longer or shorter time intervals between them. The fundamental difference is, thus, evident: Unlike space perspective, time perspective does not originate in a true perceptual process, but in a momentary cognitive re-presentation of a temporal sequence of events.

Cognitive processes are of essential importance in constructing time perspective, as mentioned above. *Cognitive representations* are for time perspective what *visual perceptions* are for space perspective. This is an important difference, indeed. Cognitive representations put us in direct contact with events, independently of their objective and real presence (Nuttin, 1984). The fact that events, integrated in a subject's time perspective, are by definition not simultaneous, does not hinder their simultaneous re-presentation. The human capacity to span time to an almost unlimited degree is of utmost importance in this context. It is a condition for the distant goal object to have an impact on present behavior and for perceiving the contingency between an action and its outcomes when a longer time interval separates them. Moreover, the flexibility of representations gives time perspective, and especially future time perspective, the versatility and creativity of the higher human cognitive functions. The only negative aspect of this representational presence is the decreasing degree of realism of events as a function of their increasing distance in time (or space).

b. Temporal signs and their origin

Our next question is about the origin of the temporal localization of mental objects within a subject's time perspective. Perceived events are present as immediate realities. They bear the temporal and spatial sign of *here* and *now*. In mental re-presentation, however, events are usually present with their characteristic spatio-temporal localization. Some events are memorized as belonging to the distant past; others, for example the publication of an article, may be represented as happening in the more or less near future. The numerous objects that I am thinking about or that, as goal objects, stimulate my actions, are all situated at different moments in time (past or future). They may co-exist in my mind with the objects of present perceptions. But each one of them has, at least implicitly, a *temporal sign*: The present act of writing is not confounded in time with the publication of the article in the future. Time perspective is built up on the basis of these temporal signs. The term *temporal sign* is used here by analogy with the local sign (*Lokalzeichen*) as used, for instance by Wundt (1910) and Lotze (1912) with regard to space references contained in sensation.

The origin of these signs is rather simple for past and present events. Their temporal signs derive from the moment of their real occurrence. The temporal sign of future events, such as the publication of an article, has two components:
1. the event is situated in the future as such, and 2. it is localized within a more or less specified zone of a particular future time period. As to the localization in the future as such, we already know that a motivational state creates a general orientation towards the future: Human needs that are cognitively processed into goal objects and behavioral projects, create a general orientation towards the future. Due to the versatility of cognitive constructs and the almost unlimited availability of symbolic objects, goal objects are mentally present before they are reached or realized at the level of reality (Nuttin, 1984). Goal objects have the temporal sign of an event 'not yet realized' or 'not yet achieved', i.e. 'belonging to the future'. Positive and negative motivational goals are situated in that future and have the general temporal sign referring to the future.

The more precise localization of an object or event in a more or less distant part of the future results from the individual's general experience with the normal course of things in his cultural environment and in the world in general. Gradually, he learns that all things have their time and take their time (duration) to be achieved. Much information is acquired by the subject observing the behavior of peers and models which he tries to imitate. Marriage, for instance, and professional retirement have a more or less exact temporal sign in the normal course of a lifetime. The intention to smoke a cigarette refers to an event in the immediate future, whereas building a house - conceived as a personal goal - is localized in a further time period. Thus, the objects an individual is concerned with, or motivated for, have their temporal sign in the same way as memorized events. The aggregate of these temporal signs, inherent in an individual's objects of concern, generates a time perspective in his behavioral functioning. It is important to realize that the objects thus temporally signed are predominantly related to the subject's motivational or affective life: personal events and objects of endeavor. Purely cognitive concepts or contents usually lack spatio-temporal localization, whereas goals to be reached or affectively loaded events do have temporal signs.

Some of the temporally localized events in a subject's mind serve as personal or social points of reference to which less important objects are anchored. In each individual life there are some key events that function as digits on a personal or social clock (in addition to one's physical and biological clock). They help the individual in the relative temporal localization of other events. Those reference marks create the temporal background against which earlier experiences or projections of future events take shape. In this context a person who intends, for instance to get married *very young* is conceived as localizing a personal event at a less distant spot in time perspective than foreseen on the social clock. This does not mean that the social clock does not work in his or her case. On the contrary, the personal impression of 'marrying very young' is based on its implicit reference to the social clock that indicates the norm.

Besides these personal localizations, there are also subjective variations and distortions in the perception of temporal distance. A marriage that is conventionally planned at a still distant period in time may subjectively be perceived by a subject as nearby. On the other hand, the sequential localization of events in time usually allows subjects to orient themselves in time in the same way as spatial points of reference improve orientations in space.

We should be aware of the fact that the temporal localization of objects and personal events in a past or future time perspective is a very delicate process: The results are usually very approximate. The process appears to be affected by age (Malrieu, 1953; Menahem, 1972). Some researchers in this field avoid the problem by making only very rough time localizations. They distinguish, for example, only between a near and a distant future or past. We tried to solve this problem in our own research method by introducing a so-called 'objective' or 'mean' time localization, instead of asking the subject himself for his personal temporal impressions. The justification of this method will be given below (cf. p. 68).

c. The presence of objects in time perspective

Our conceptualization of time perspective implies that events with their temporal sign are 'present' in a subject's time perspective in the same way as objects located in space are present in his spatial perspective. Indeed, perspective is created by the actual presence of objects at different distances. But what do we mean by 'being present' in a future time perspective?

Although an individual can only think of one situation or one category of objects at a time, his time perspective is not constituted by the localization of that element only. Perspective usually implies the presence of more than one object, although not necessarily in a very focused way. A comparison with the spatial dimension will clarify what we understand by the mental presence of several objects. A person usually spends his life and moves in a certain 'field' of which the *action radius* can be measured. Most of his behavior 'radiates' in that sector. One person may live and work within the bounderies of his little provincial town without leaving it. Another may travel very often by car, train, or plane throughout his country. A third person may have the whole world as his sphere of action. Each of them, however, can only be at one place at one time. It may happen that the world traveler is home in his town and that our 'villager' visits his son in a city at the other side of the country. But, the usual action radius of these persons differs considerably. The same holds for time perspective. If we take a sample of the events and goal objects an individual (or homogeneous group of subjects) is occupied with, it is very likely that for some persons (or groups) those objects will span a much longer time period than for some others. At each moment in time, the subject's attention is focused on only one event, but many other objects or events are present in a habitual, latent, or virtual way. This virtual presence will manifest itself as soon as the situation requires it; for instance, when the subject is requested to communicate the objects of his current concern. Those objects can, therefore, be said to occupy the subject's mind in a *habitual* or *virtual* way. The objects and events the subject is concerned with in this way and which motivate and codetermine his actions may change as a function of life conditions or life situations. In dangerous and stressful situations, time perspective may be limited to very immediate goals or objects, while in periods of calm reflection, very long past and future perspectives may unfold.

A specific problem to be solved in each research project is to outline the situational and behavioral unit in which the time perspective is to be investigated. Without going into this problem in detail (see Barker, 1963) we may say, first, that the goal of the research should determine how broad the temporal period used as a unit should be; second, that we should consider molar behavioral activities that have a special meaning within the subject's life cycle. For example, the future time perspective of young recruits who start their one-year military service immediately after graduating from college can be compared with that of an equivalent group of graduates who are dispensed from military service and immediately start the first year of their professional career. What is the time perspective of people who are hospitalized or imprisoned, as compared to an equivalent group of active professionals? What is the impact of stress during the final hours before a decisive examination, or during the period following an important (experimental or real-life) failure? Each of these life conditions may be studied as behavioral units in which the temporal localization of motivational objects and concerns virtually occupying the subject's mind can be studied. As said before, we accept that an object *virtually* participates in a subject's time perspective when it effectively comes up in his mind when a situational factor - such as a researcher's instruction - invites him to express the objects of his current concern and motivation.

In summary, an individual's time perspective is conceived as the configuration of temporally localized objects that virtually occupy his mind in a certain situation. It is not limited to the single object that the subject has in mind at a given moment, in the same sense as space perspective is not limited to the object actually focused on by the perceiving subject. The peripheral and intermediate context is essential for perspective.

Two points of view can be taken in defining time perspective: the subjective view of the individual on the one hand, and the arrangement or sequence of objects on the other. From the subject's viewpoint, time perspective is the temporal zone to which his mental view virtually extends itself when considering the objects and conscious determinants of his behavior. From the 'objective' point of view, time perspective is the totality of objects located within a more or less extended temporal zone insofar as they are virtually present to the subject in relation to his behavior.

The presence of a particular time perspective creates in the subject a habitual openness towards the future or the past. Its absence means confinement within a continuous present. The virtually present object becomes effectively present as soon as it is aroused by the situation or by a relevant motivation. For example, the reward that is implicitly expected by a child at the moment his parents will come home, is part of his future time perspective. It has a temporal sign of 'tonight', i.e. the moment that the parents are expected to come home. The object *reward* is virtually present in the child's mind and it will now and then be consciously (effectively) present during the day when, for example, he perceives some objects (a picture of the parents, the candy box, etc.). Due to this virtual mental presence of the reward, the child will behave differently in some situations. The object - i.e. the future reward - is part of his behavioral world in the sense that it is a kind of goal object to be reached. In general, *the objects of time perspective are among the determinants that regulate behavior.*

In addition, the extension of the future time perspective has an important role in the elaboration of behavioral plans and projects. The individual who looks far ahead formulates long-term projects and may find more means to realize them. If means are lacking in the present situation, he may search for them in a more distant future, as will be shown below in the example of the globe-trotter with a long radius of action. Therefore, a long, realistic, and active future time perspective is important in planning and realizing behavioral projects, since almost all important achievements require coordinated and long term means-end structures.

Note

The reader will have noted that in our conceptualization, *future* time perspective is formed by the more or less distant goal objects that are processed by an individual. On the other hand, we claim that a long future time perspective is a prerequisite for the elaboration of long term projects. There is, however, no real contradiction between these two statements, if one accepts the existence of reciprocal influences within developmental processes. Let us take up again the analogy with the spatial action radius (cf. p. 20). The man who never leaves his hometown will hardly consider a motivational goal or a means that requires him to travel to another country, while such a means-end structure will easily be available in the globe-trotter's motivational construct. In the same sense, a long future time perspective is required for the elaboration of long term projects. But it is equally true that the same long-term projects contribute to the creation of long future perspectives, just as the world trips of the globe-trotter gradually create his action radius. Perspective is the result of a progressively developing process. To the extent that an individual - imitating social models or reinforced by successful past experiences - constructs new goals, he progressively goes beyond his previous temporal horizon. Formulating new future projects is facilitated by doing so. It is obvious that an intellectual concept of the unlimited chronological future is also required for such a development, although it is not sufficient to that purpose. To be intellectually aware of the world space is a necessary, but not a sufficient, condition for the villager to plan a trip to another continent.

IV. AN OPERATIONAL DEFINITION: MEASURING TIME PERSPECTIVE

In our definition of time perspective we have stressed its 'material' or 'objective' component, i.e. the past and future objects constituting the content of time perspective. This is important, because time perspective is not a pre-existing 'empty space', unlike the abstract notion of time; it cannot be conceived independently of its content, since its temporally localized objects are the basis of its temporal zones and their density and extension. On the other hand, concrete goals and means - as well as memories - do not really exist outside a certain time perspective, since they have a temporal sign inherent to their content. Content is an essential element of time perspective, and the temporal dimension is an essential element of concrete goals and memories. Therefore, both aspects - content and temporal dimension - have to be taken into account when studying past or future time perspective.

The most elementary structure within the stream of time is obtained by dividing it into two parts: the past and the future with the present moment as its point of reference. Objects are easily located in one of these three sectors. The explicit temporal signs of some key events are further used as points of reference for a more precise localization of other events. Our problem, then, is to know how a subject's *temporal perspective can be measured?* How can it be operationalized? At this point it should be remembered that time perspective in its proper meaning is to be distinguished from *time orientation* and *time attitude*, as noted above (cf. p. 11). In our opinion, the measurement of time perspective proper is related to the following aspects: 1. its *extension* or length or depth; 2. the *density* with which the objects are distributed within the different periods (past and future); 3. the *degree of structuration* among those dispersed objects, i.e. the presence or absence of ties between objects or groups of objects (for example, means-end relations as opposed to a mere juxtaposition of objects); 4. the degree of *vividness* and *realism* with which the objects are perceived by the subject, as a function of their distance in time. All these aspects are related to the temporal localization of the objects and to the links that tie them together in time perspective. Time orientation and time attitude, on the contrary, are different research topics. In the method proposed here, the stress is placed on the extension of time perspective and the density of objects within different temporal zones.

Measuring time perspective is generally based on two types of methods. The first can be called 'object methods'. The basic data are the past and future events or objects which, at a certain moment in time, virtually occupy a subject's mind and are of some concern to him, as explained above. The measurement technique that we developed and that will be described in detail further in this book can be summarized as follows. A sample of the individual's motivational objects and goals is first collected. This is done by means of a verbal technique: our *Motivational Induction Method* (cf. Chapter 2). Second, the motivational objects thus collected are localized in time on the basis of a more or less differentiated temporal scale: Trained judges give each motivational object its temporal sign on the basis of a previously elaborated coding system (our *Time Code*, see Chapter 3). The principle underlying this procedure is that motivational objects are mentally localized by the subjects in a time period where their realization is most normally or probably to be expected.

Such a localization in time allows us to measure in a more

objective and comparable way the total extension of a subject's time perspective and the more or less dense distribution of the objects in each of the different time periods. Such an analysis results in *a time perspective profile* indicating the relative number of objects in each time period. The mean temporal distance of the goal objects of an individual or a group can be calculated and compared between subjects or between groups in different experimental conditions, as will be explained later. The effects of experimental or differential variables on the extension and density of the time perspective are reflected in this kind of measurements. Other aspects, such as the degree of structuration and the level of realism, are analyzed by using additional techniques.

The second and more 'formal' type of measures of time perspective uses creative expressions such as stories or drawings (circles or lines, for instance). Formal characteristics of these creative expressions, such as duration, relative length or size, number of verbs in the past or future tense, are *interpreted* as symptomatic for the extension of the subject's time perspective. Most of these techniques are more or less projective (Tell me a Story Technique; Story Completion Technique; T.A.T.stories; Circle Test; Line Test, etc.). To the extent that they are really projective techniques they may be able to give some indirect information about motivational and affective mental contents (cf. Lens, 1974).

A special advantage of our *object method* as presented in the following chapters should be mentioned here. By taking an individual's goals and objects of concern as the basis for measuring his future time perspective, we are able to introduce an interesting differentiation within time perspective depending on the subject's fields of activity. In fact, a subject may have some specific objects of concern related to one or other of his special roles or qualifications in social life. Our method allows us to investigate a subject's concerns and his time perspective insofar as he/she is, for example, a housewife, an American, a Jew, a Huron-indian, a female, a Christian, a Communist, etc. Differences in time perspective can be studied as a function of such different roles (cf. infra, p. 63).

Another type of differentiation can be introduced by measuring the temporal localization of objects of an individual's thoughts as compared to the objects of his fears and of his positive motivation or desires. Thought contents may refer to the past as well as to the future, whereas motivational objects are normally localized in the near or distant future. Objects of fear may have another temporal dimension than desires, aspirations, and behavioral projects. Moreover, goal objects constitute a more *active* time perspective than thought contents do (a *cognitive* time perspective as opposed to an *active* or motivational one). Besides the study of time perspective in different groups of subjects (age groups, prisoners, socio-economic groups, etc.), different time perspectives could also be studied within one group as a function of different personality aspects. Thus, it appears that talking of time perspectives - in the plural - refers not only to the past and future time perspective, but also to the time perspective involved in different behavioral areas and personality aspects. It is not unlikely that these different time perspectives do not highly correlate. The term *global* time perspective may be used with regard to measures that imply all kinds of a subject's motivational concerns, as distinguished from more *specific* perspectives. The *total* time perspective, then, refers to the total extension of a subject's time perspective, going from his past to his future horizon. As to the term *temporal horizon* itself, it is often used for referring to time perspective itself (see among others, Fraisse, 1963). We prefer to reserve the term temporal or time horizon for referring to the most distant area of future time perspective and to the objects situated in that area. The future time perspective itself encompasses the whole range of temporal zones. This restricted meaning of temporal horizon is in accordance with the spatial area where the horizon also refers to the most distant zone in the perceptual field.

V. ADDITIONAL ASPECTS OF TIME PERSPECTIVE

a. An active time perspective

In defining time perspective we referred to objects that are present on the level of mental representation. It is obvious, however, that a large number of objects of thought and representation do not have a temporal sign: concepts, notions, the meaning of a word once learned but now understood independently of any temporal context, etc. It is obvious that the study of time perspective is concerned only with objects that can be situated in time. Some authors even want to limit time perspective to those objects that, effectively, have some influence on the subject's overt behavior: a goal object, for example, that regulates action and for which one would be willing to suffer frustrations (Leshan, 1952; Heimberg, 1963; Lessing, 1972). We assume that, to a certain extent, this is the case for what we call motivational objects. On the other hand, as far as a subject dreams about non-motivational objects and "castles in Spain", he has a *cognitive* or *imaginative* time perspective. Kastenbaum's (1963) distinction between *personal* and *cognitive futurity* is analogous to our distinction between an *active or motivational time perspective* on the one hand, and a *cognitive or imaginative time perspective* on the other.

The distinction between these two types of time perspective is not always easy to make because most studies only use verbal measures. However, measures based on a "Tell-me-a-story-technique" are less likely to measure an active time perspective than more direct measures that are based on samples of motivational goals or personally important events. Measures of the second type are used, among others, by Brim and Forer (1956), Rizzo (1967-1968), Blatt and Quinlan (1967), Lessing (1972), and in our research with the MIM. The research of Raynor (1969; 1974), Gjesme (1974; 1976) and De Volder and Lens (1982) also concerns an active future time perspective.

With relation to the concept of *active* time perspective, an important remark should be made. There is a danger in limiting the behavioral effects of time perspective to overt behavior only. Certain memories and anticipated motivational goals may arouse no overt behavior at all, but they may have a debilitating or facilitating effect on *covert* psychological functioning. Objects of unrealistic aspirations or past frustrations may strongly affect cognitive and emotional processes and even the well-being of an individual in general when they are associated with real motivational tendencies. That is why, following our conceptualization of behavior (Nuttin, 1984, Chapter II), those objects belong to an active time perspective.

Although the objects of personal tendencies are the most important for the content of a person's time perspective, the anticipation or memory of someone else's action and its outcomes as well as impersonal events, are part of it. To the extent that they are related to the person's individual interests, they are part of his active time perspective. On the other hand, events that are objects of a purely cognitive time perspective are by definition neutral (not active). Therefore, the cognitive time perspective of a historian studying the 14th century, or a futurologist interested in the 21st century, is to be distinguished from the individual's active and personal time perspective.

b. Realism of time perspective: temporal integration and temporal competence

The degree of realism of the objects constituting an active time perspective is an important variable affecting its behavioral effects. This effect decreases as a function of the progressively decreasing degree of realism of distant objects (in space and time). In fact, there is a temporal gradient - as well as a spatial gradient - that affects the strength of a goal-oriented tendency. Objects that are more distant in time and space have a lower reality value and less behavioral impact. In his first publication on the temporal dimension, Lewin (1931) opposed the maximum degree of realism of a *perceived* object to the less real nature of objects in mental *representations* (such as in time perspective). But even among objects of representation, different degrees of 'reality' and, hence, of psychological impact, have to be distinguished. Thus, objects beyond an individual's habitual temporal horizon appear as less real to him. Therefore, a short time perspective is an additional variable decreasing the degree of psychological reality of distant objects. Moreover, structural factors, such as the presence or absence of perceived causal or instrumental relationship between the objects of one's time perspective, are likely to influence the subjective degree of realism of future objects.

Generally speaking, people tend to perceive a temporal sequence of events as causally related to each other. An event following a personal act, in particular, tends to be perceived as a consequence of that act. Inversely, a goal object is usually conceived as to be realized by relevant activity. This problem has been studied in different contexts.

Studies on attitudes towards the future show that some subjects see the future as determined by chance. Others view it as mostly dependent on their own actions; they see it as something that can be more or less controlled and realized by themselves. In that case, their own activity has a higher degree of perceived instrumentality. Research has shown, for instance, that achievement motivated students who perceive their studies as instrumental to their *future* career are more study-motivated than those who do not perceive this instrumental relationship. Another type of research in this field concerns Rotter's (1966) concept of *locus of control*. Some people perceive what happens to them, and certainly the outcomes of their actions, as determined by external factors (*external attribution*), whereas others attribute those events and outcomes to internal factors such as capacity and effort (see also Weiner, 1972; 1974; 1980).

In the two types of research mentioned we find two fundamental requirements for a future time perspective to affect present behavior: First, a certain degree of *temporal integration* is needed, so that the future (the career, for instance) is seen in an active continuity with the present and the past (one's studies, for instance); and secondly, a disposition to make *internal causal attributions* recognizing the role of personal action in achieved outcomes.

I hypothesize that as soon as an individual starts *to work for* a distant goal, a causal relation is established between the present activity and the goal, so that the degree of reality of that goal object progressively increases. The means-end structure that bridges the time interval fills up the distance between the present activity and the goal object. An empty time interval, on the contrary, is unrelated to the individual's actions and increases the psychological distance to the goal object. According to this hypothesis, an active and realistic time perspective is largely dependent on subjective perceptions (perceived instrumentality, for instance) and on the existence of means-end structures (temporal integration). A subject with a high degree of such *time integration* can be said to be *time competent*.

The notion of *time competence,* which was introduced by Shostrom (1963; 1968), as well as the ideas of some therapists stressing the negative effect of time perspective, are to be situated in this context.

In the last few years, it is no longer generally accepted that a long time perspective favors efficient and goal-directed behavior. Some psychotherapeutic schools assert that it is much more efficient to live completely in the present without any time perspective at all. People for whom the past or the future somehow play a role in their behavior - people with a time perspective - are said to be unable to actualize their potentialities (self-actualization). Certain conceptions of Maslow, Rogers, Perls, and of Gestalttherapists are related to this way of thinking. It deserves our attention to the extent that it can help us in specifying the notion of time perspective and its real importance. Indeed, it is possible to have one's mind completely occupied with past and future events so that one can no longer take advantage of the present situation as a source of behavioral action. Clinical psychologists and psychiatrists often have to deal with such persons as patients. But the scientific analysis of time perspective has to be situated in a different context. For a normal and active person, the future is not an escape from present reality; it is the world of his goal objects which direct and coordinate his present activities. Moreover, as I noted above, the more important human achievements take time: A rather long temporal span is required to conceive important behavioral projects and means-end structures leading to important final goals. From the very beginning, behavior has to be regulated by these final and distant goals that are present on the level of cognitive functioning. On the other hand, the lack of any future time perspective is characteristic for people interested only in the immediate satisfaction of physiological needs, as is the case with children and in some primitive societies. The same may happen in stressful situations.

From what has been said, it appears that some misunderstanding, or at least some ambiguity, exists in the conceptualization of past and future events constituting an individual's time perspective. One important conclusion should be that their role is complex and multiple. Objects of future time perspective may have either a substitutive or a preparatory role in behavior, depending on their being pure fantasies (wishful thinking) or real goal objects. A fundamental variable determining their behavioral impact is the perception of a causal or instrumental link between the present moment and the representational objects of time perspective , as noted above. The student who sees his studies as instrumental for his future career has a realistic future time perspective and an integrated perception of the present and the future (*temporal integration* and *competence*). The individual who 'dreams' of an important career, but without actively preparing it, may become the patient of a psychotherapist who, therefore, will tend to devaluate the importance of an outlook on the future and to accentuate the importance of present activity. Perceiving the link between present activities and events that have their place in time (temporal integration) is a condition for a *realistic* time perspective. Such an *active* time perspective is an important variable in the study of behavior and its motivation.

Some psychotherapists who emphasize the importance of being oriented towards the present - such as Shostrom himself - have a more balanced point of view. They recognize that there are pathological types of orientation towards the present, as there are pathological orientations towards the past and the future. The individual who lives in the present without any realistic future-oriented projects instigating and coordinating his actions is pathologically oriented towards the present. The task of a psychotherapist in such cases is to direct the patient towards the future, while showing the links between present behavior, distant goals and past experiences. The individual who has *time competence* perceives the temporal continuity and integration of events. His actions happen in the present but are stimulated and guided by future goal objects. To the extent that goal objects lose their preparatory and coordinating function, the cognitive representation becomes a substitute for real action. It does no longer refer to the future; it becomes an atemporal fantasy.

c. Experiential learning and time perspective

Many anthropological and sociological observations reveal a very restricted future time perspective among people living in unfavorable cultural or socio-economic conditions. In the same way as it may be desirable for neurotic patients to abandon the future and to live in the present, it may be realistic for members of some socio-cultural groups to make no plans or behavioral projects for the future. Unforseeable external conditions, such as unstable political situations, may strongly affect the predictability of their personal future. Unlike for normal and active men, their future is not a field of creative possibilities, but an object of absolute uncertainty; it is passively waited for. In other cultures, a fatalistic philosophy of, or an obsession with the factor *chance* or *good fortune*, produces an absence of planned actions. For these people, developing projects or even working for distant goals would be a sign of naïveté and irrealism (Cottle & Klineberg, 1974).

Hence, we should recognize the importance of life conditions and personal experiences in the elaboration of time perspective as well as in abandoning any future project. Research on delay of gratification shows the strongly different reactions of children who repeatedly experienced unkept promises in comparison with children not knowing such deceptions. Bochner and David (1968) found that children of Australian aborigines who preferred a smaller immediate reward rather than a greater but deferred gratification, had significantly higher I.Q.'s than their peers, and this at each age level. In this culture, intelligent children learn to distrust the unreliable future. Battle and Rotter (1963) report the same trend for locus of control. Among black North American children living in unfavorable socio-economic situations, those who believe in an external locus of control generally have a higher I.Q. than the subgroup with an internal locus of control. The effect of earlier experiences may also partly explain the restricted future time perspective usually found in some social classes of our Western population. These findings raise the important problem of the extent to which time perspective can be acquired and developed through education. Until now, few researchers have systematically investigated how to influence the extension of time perspective. Ricks, Umbarger and Mack (1964) provide an interesting example of such an approach.

VI. THE IMPACT OF FUTURE TIME PERSPECTIVE

The impact of time perspective on present behavior is, no doubt, the major problem for theory and research in the field of psychological time. In view of our earlier definition, two aspects of time perspective have to be investigated. From the point of view of its content, time perspective consists of mentally represented *objects* localized in different time periods. Therefore, we have labeled this aspect the 'objective' concept of time perspective. In the past time perspective, these objects belong to the field of memory, whereas an active future time perspective is mainly filled up with short- and long-term goals and means-end structures (i.e. behavioral projects and plans). As to the second aspect of time perspective, it consists of its temporal dimension as such, especially its extension, structure, and density. Limiting ourselves to the *future*, the processes involved in the impact of both the 'objective' and the temporal aspects will now be briefly examined.

a. The content of future time perspective:
  Its impact on behavior

As just said, the 'objective' or content-aspect of future time *perspective* consists of the goal objects and means-end structures that an individual has virtually in mind when behaving in a present situation. Therefore, the basic processes via which an active time perspective influences behavior are identical to those by which goals and means-end structures regulate action. In fact, goals on the one hand, and the future time perspective in which they are embedded on the other, should not be separated: Goals are inherently localized in that perspective, and the perspective itself is built up on the basis of these motivational objects. The impact of goals cannot be adequately studied outside the temporal dimension affecting their reality.

It is not my purpose here to describe the processes via which goals and behavioral projects or plans activate and regulate behavior. This I have tried to do in another book (Nuttin, 1984). I only intend to emphasize the *motivational nature* of the processses involved. Under the influence of cybernetic models on the one hand, and cognitive psychology on the other, motivational processes tend to be eliminated from the impact of goals as contained in future time perspective. The impact of goals is often conceived in terms of expectancy of behavioral outcomes or of feedback of previous outcomes, without referring to the motivational processes underlying the influence of outcomes. In other cases, goals are conceptualized as associations of behavioral objects with positive or negative affect (pleasure or displeasure) without referring to the motivational basis of that affect itself. Let us consider briefly these interpretations.

The regulating impact of outcomes is due to their being either in congruence or in dissonance with motivational tendencies or standards set up by the subject. In other words, behavior is not regulated by its outcome as such, but by the subject's preference for some outcomes as compared to others. This preferential directedness towards some categories of outcomes is precisely the effect of motivation. In cybernetic models, and in self-regulating devices such as thermostats, the program and standards that are introduced are the results of what a *motivated subject* intends to obtain.

As to the association scheme, it is not sufficient to say that an object becomes a goal by virtue of its association with a pleasurable or gratifying affect. As we have shown elsewhere (Nuttin, 1983), conative and sensorial affects imply a dynamic fit; affect is the response of a dynamically or preferentially oriented individual when encountering fitting or nonfitting outcomes or objects. This does not exclude that, in turn, the positive or negative affective response itself becomes motivating and regulating in further behavior.

As to the expectancy-value theory of motivation, the motivating impact of goal objects is explicitly recognized by mentioning the value factor. The expectancy process however, is of a cognitive nature and relates to the estimated probability with which an act is expected to lead the subject to the goal. In our opinion, the process via which expectancy co-regulates the motivational intensity of the instrumental act is to be conceived as follows. Instrumental acts derive their dynamics from the subject's motivation towards the goal (the *valence* of the goal object). This goal object is a concretization of the subject's behavioral needs that are at stake; its valence depends on the intensity of the subject's need(s) and the need-satisfying properties as perceived in the specific goal object. The dynamics thus invested in the goal object are channeled towards the instrumental act to the extent that the act is perceived as a good path or instrument towards the goal (perceived instrumentality). This means that the intensity of instrumental motivation is regulated, not only by the intensity of the subject's dynamics towards the goal, but also by the channeling process regulating the amount of motivation transferred(5).

Among the other factors regulating the motivational intensity, we note the number and efficiency of other instrumental acts available for reaching the same goal. The attractiveness or the aversive character of the instrumental act (and its anticipated outcome) as such is also to be added to, or subtracted from, the amount of motivation derived from the goal object. Moreover, in many cases instrumental acts are motivated by several goals and subgoals. Thus, in making a choice between different performances in a competition, a subject may be motivated not only by the money reward (the main goal), but also by his desire to show his high degree of competence. Therefore, it can be predicted that the performance chosen will not be the easiest one available, i.e. the instrumental act with the highest probability of success; in fact, the preference will go to a more difficult performance showing the subject's competence (second goal) without excessive risk of losing the prize (cf. research on achievement motivation). Thus, the two goals will be reached optimally. This combined action of goals explains why some formula's for predicting the subject's choice do not always hold, as shown by Shapira (1976).

From what has been said it appears that the role of expectancy of outcomes in regulating motivation is a complex one, depending on the goals set and on the means available. In the cases just described, expectancy is a cognitive process of perceiving the probability of an instrumental outcome, i.e. the cognitive anticipation that the act will lead to the goal(s) set (perceived instrumentality). In this cognitive sense, expectancy has no motivational impact of its own, but it regulates the amount of motivation that is channeled from the goal object to the instrumental act as mentioned above. When it is said, for instance, that not the goal, but *its anticipation* 'causes' behavior, i.e. activates and directs it, this ambiguous expression should be understood in the following context. A goal only exists on the cognitive or mental level of behavioral functioning. That means that it only exists insofar as cognized; and a cognized goal is an anticipated *goal,* since goals are situated in the future (as still to be realized). Therefore, anticipating the goal is the only way to actualize it or to bring it to psychological existence. Thus, only the actualized or anticipated goal psychologically exists and is able to exert its motivating effect. This motivating effect of the goal ultimately resides in the behavioral need that initiated the cognitive-dynamic searching, producing the goal. Therefore, the motivating power of the anticipated (actualized) goal stems from the behavioral dynamics that are concretized in the goal, and not from its anticipation.

It must be added that, usually, the terms *expectancy* or *expectation* do not refer to purely cognitive anticipation, but may have some motivational implication, as we have already emphasized in another context (Nuttin, 1953, p. 149; see also Heckhausen, 1977). When a child is said not to come up to the expectations of his parents, something more motivational than anticipation of probability of outcomes is implied: the parents' expectations imply their desires and hopes, and hope is a mainly motivational concept (6).

Another concept often used as a substitute for motivation is *discrepancy*. The term itself implies a psychological 'distance' between two situations, or between the subject and an object. A psychological distance, however, is motivating only to the extent that a subject is already motivated to reduce that distance. In other words, an object is, psychologically speaking, *distant* or *discrepant* inasfar as a subject is motivated to be "close" to it (i.e. "to have it") or to be congruent to it. Once more, the motivational effect of discrepancy refers to underlying dynamics.

In summary, the process through which the content of a subject's future time has an impact on present behavior is to be identified with the motivational process by which goals and means-end structures regulate behavior. Other concepts such as *outcome, affect, expectancy* and *discrepancy* cannot be substituted for motivational processes; they play a role in regulating behavior only to the extent that motivation is implied.

b. *The impact of the temporal extension of time perspective*

Our second point of view in studying the impact of future time perspective is related to its specifically temporal dimension, i.e. the extension of time over which the objects are spread, their density, structure, and reality character. To what extent do these time-related features of goals and plans influence their impact on the motivational intensity of present instrumental behavior? First, it should be emphasized that an extensive time perspective contributes to the subject's setting of distant goals and to his elaborating long-term behavioral projects. As shown above (cf. p. 20-22), the subject with a short time perspective is handicapped in looking into the distant future for means and ends to satisfy his needs. In the same way as the villager who never left his town will hardly think of means and ends available only through traveling abroad, a subject without a future perspective will be limited in his goal setting and plan making to a few hours or days.

Thus, an extensive time perspective stimulates long-term goal setting; it provides subjects with more 'things to do' and long-term projects to work at. Time perspective seems, therefore, to be at the basis of a richer variety of purposive activity. It would be interesting to investigate to what extent a more extensive, a more dense and more structured future time perspective correlates with a great variety of activities, either on the level of overt behavior or on the level of creative imagining. As to the problem of the impact of distant goals, it is well known, and generally accepted, that the impact of temporally and spatially distant objects decreases as a function of their distance. The same holds for goal objects. Distance diminishes the psychological reality and, hence, the impact of objects and events. It is an intriguing problem, for instance, that people who seem to be very concerned with their health cannot be motivated to refrain from doing *now* something that will certainly do them serious harm in the long run. The first process intervening in the motivational impact of future time perspective is related to this temporal distance issue. People with an extensive future time perspective are not only likely to develop a greater variety of distant goals and projects as just mentioned, but these goals do not lose their reality character and, hence, their impact to the same extent as is the case with subjects for whom distant goals lay beyond their usual time horizon. Thus, it is assumed that distant goals are more motivating in people with a longer future time perspective (see De Volder & Lens, 1982).

A second process affecting the impact of distant goals on motivation for present behavior is related to the means-end structure of behavior. As mentioned above, it appears that people with an extended and structured time perspective are more likely to perceive the instrumental relationship between present behavior and distant goals. It has been shown that achievement-motivated students perceiving the instrumental relation between their present study behavior and the distant career are more motivated for the present study than those who do not perceive this relation. In accordance with the process described above with regard to the expectancy x value theory, it may be assumed that the motivation for the goal object (the career) is more fully channeled to the instrumental act (present study) when the instrumental relationship is actually perceived.

Summarizing the second point in our discussion of the

impact of the future time perspective, we can agree with De Volder and Lens(1982) when they state: "The *dynamic* aspect of future time perspective is formed by the disposition to ascribe high valence to goals in the distant future. The *cognitive* aspect is formed by the disposition to grasp the long-term consequences of actual behavior, as reflected in the concept of instrumental value of a behavioral act". With regard to the field of applied psychology, it may be important to notice that the total amount of motivation invested in the instrumental act - e.g. in study behavior - can still be negative in subjects with extensive time perspective. This will be the case, for example, when the aversive character of the instrumental act as such (study behavior) is stronger than the positive goal-motivation channeled to the instrumental activity. This will even be more the case when the goal itself - the career, for instance - is imposed from the outside, without intrinsic motivation on the part of the subject. In these cases, the instrumental act will not be performed in the sense that the child will not really study, although it may attend classes for other motives.

A final remark with regard to motivation and future time perspective: In studying motivation, psychologists usually limit themselves to the impact of only one motive on behavior. In real-life situations, however, a subject's motivational field is an hierarchical structure of many goals and projects virtually present in his future time perspective. Our method for studying time perspective is based on this idea of the mental presence of several goals, aspirations and fears. This emphasis on the whole range of the subject's goals and plans as situated in their temporal and thematic context is the real importance of the concept of future time perspective in the study of motivated behavior. Therefore, the motivational content of time perspective has been emphasized in our approach. In this sense, the investigation of time perspective should contribute to the study of motivation in its structural and temporal context.

CONCLUSION

We held that the term *time perspective* should be used in its strict sense in research, distinguishing it from *time attitude* and *time orientation*. General future-orientedness in behavior cannot be derived from memory of past experiences; it is a new dimension introduced by states of need, i.e. by motivation: The future is motivation's 'space'. Various degrees of depth (extension) in future orientedness - i.e. future time perspective - are created by the specific objects in which needs manifest themselves. By analogy with space perspective, the *objects* 'localized' in time are essential in constituting a perspective. Time perspective, indeed, is neither an empty 'space' nor an idle conceptual frame. While cognitively 'present' on the perceptual level in space perspective, the objects constituting the content of time perspective are 're-presented' on the symbolic or representational level of cognitive functioning. In past time perspective, the underlying objects arise from the active memory, whereas the content of future time perspective stems from virtually present motivational objects, i.e. goals and behavioral projects still to be realized, objects and events one is looking forward to or afraid of. The 'temporal signs' affecting these objects - i.e. localizing them in future time - are gradually built up on the basis of cognitive experience in daily life. Thus, by conceiving a goal object in the frame of a need to be satisfied, a certain future extension is introduced on the representational level of behavioral activity. On that representational level, past and future objects are actually present. This mental presence of future goal objects is required to construct and regulate the means-end structures implied in purposive behavior.

As to the processes via which future time perspective influences present behavior, the distinction made between its motivational content and the temporal dimension as such is to be taken into account. The impact of more or less distant goal objects is to be accounted for in terms of motivational processes, which cannot be replaced by cognitive processes such as *expectancy* or *anticipation* of future outcomes. As to the decreasing impact of goals as a function of their distance in time, the positive influence of an extended time perspective on the valence of distant goals and on the perceived instrumentality of present behavior for obtaining these goals has been emphasized.

With regard to the methods for measuring the extension of future time perspective, it follows from our operational definition that - in a first step - a sample of current motivational objects is to be obtained and that - in a second step - the average temporal localization of these objects is to be coded. The following chapter of this book will be mainly devoted to the description of this two-step method. A brief report of several researches carried out with that method can be found in Nuttin (1980, Chapter 5).

More than once in this chapter on theory, the importance of an extensive future time perspective for constructing long-term behavioral projects and, hence, for accomplishing important achievements has been emphasized. Generally speaking, one could say that the future is the 'mental space' in which human needs are cognitively processed into long-term goals and behavioral projects. In this sense, the mental construct called the 'future' is the building site of constructive behavior and human progress. This active and realistic future time perspective is combined with time integration and time competence.

Parallel to the three dimensions of time - the present, the future, and the past - three cognitive functions build up our three-dimensional behavioral world: While perception establishes man in his present situation, selective memory reconstructs his experiential past, and the future comes into existence on the level of dynamic and creative representation and imagination.

Therefore, the future time perspective is an essential dimension of a person's behavioral world and of his motivated purposive action.

NOTES

(1) When a *perceived* object is taken as a goal object, the act of taking it as a goal, i.e. as an object to be obtained, is a cognitive-dynamic activity on the representational level of functioning.

(2) For this, we refer to Fraisse's chapter on 'temporal horizon' (Fraisse, 1963) and to the review articles by Wallace and Rabin (1960) and Mönks (1967). We do not discuss either the many studies on time perception, time estimation, time sense, and other aspects of the psychology of time.

(3) Early in this century other authors studied the effect of the future on memory. Aall (1912; 1913) used the concept 'time perspective' (*Zeitperspektive*) in its present

meaning as the anticipation of a future event or goal object. He studied its effect on the length of the retention period. In his doctoral dissertation, Nuttin (1941) analysed the effect of an 'open task' (open to the future) on learning and the law of effect. He found that a rewarded response is better remembered when the individual expects that the response will be useful in the future (Nuttin, 1953, chapt. VI & VII; Nuttin & Greenwald, 1968, chapters 5 & 6). The numerous studies on the effect of intentions on learning are situated in the same context. We refer to Ryan (1970; 1981) for a detailed discussion of that research.

(4) The many non-technical meanings of the term *perspective* are not of interest to us; they may, however, have contributed to the confusion.

(5) A large number of instrumental paths or behavioral projects leading to the same goal may be constructed by the subject and, thus, motivated through derivation of the dynamics toward the goal. It is interesting to note that it is essentially the same motivation for the same goal that remains active in all these instrumental acts and plans. Additionally, each instrumental act may have a certain amount of attractiveness or aversiveness of its own.

(6) The regulating impact of expectancy on the intensity of motivation, as discussed here, refers to the motivation for an instrumental act. As contrasted with the instrumental act, the intensity of a subject's motivation for the goal object itself does not depend on the expected probability of reaching that goal. When, for instance, after several unsuccessful trials, the inaccessibility of a goal is eventually perceived by the subject, his overt behavior towards that goal may be stopped, but the enduring emotional repercussion of the frustration may still testify to the intensity of the motivation, although the probability of reaching the goal is zero.

CHAPTER II

THE MOTIVATIONAL INDUCTION METHOD
(MIM)

Our operational definition of time perspective, as described in the preceding chapter, advocates a two-steps measurement technique. First, a sample of objects that interest or motivate subjects has to be collected; second, these objects are to be situated on a temporal scale in order to measure the extension and other aspects of a subject's time perspective. Only the method for measuring *future* time perspective will be described here.

The proposed method is based on two principles:

1. Future time perspective is constituted by motivational objects that exist virtually on the level of a subject's mental representation; it therefore should be measured by identifying the temporal localization of these objects (this is what we call the 'object approach');

2. In measuring a subject's time perspective it is highly desirable not to suggest to the subject any specific time period. In fact, any instruction asking the subject what he would like to do or to be in, say, five or ten years from now, may evoke objects that might otherwise not occur to him and, therefore, do not really exist in his behavioral world. As to the use of projective techniques for measuring time perspective, the temporal meaning of their data is difficult to interpret with scientific objectivity.

Unfortunately, the method proposed here also has its negative aspects and difficulties. Collecting representative samples of motivational objects and localizing them in time is a complex enterprise. It takes more time than simply asking subjects to tell a story or to draw lines representing the length of their past and future, and it is easier to ask subjects to imagine what they will be doing in five or ten years from now, than to situate in time their spontaneously expressed motivational objects. However, our theoretical viewpoint strongly recommends the more laborious and time-consuming method proposed here. Its first step will be briefly described in the present chapter, and the general background of the temporal coding and measurement techniques will be explained in the two following chapters. Two *Manuals* at the end of this book provide the researcher with the necessary details for applying the method.

In a first step, then, a sample of the motivational objects of a group of subjects is collected with our *Motivational Induction Method (MIM)*. This method is of the sentence completion type but, unlike most such instruments, it is not used as a projective technique. The sentence beginnings, called motivational *inducers*, are formulated in the first person singular and induce the subject to express - usually in writing - a number of concrete objects he desires, pursues, fears, or tries to avoid (positive and negative goal objects). It is the task of the researcher to analyze these data and to identify the more general motivational categories to which they belong, in the same way as he tries to identify the meaning of a subject's behavior patterns mentioned in response to a personality questionnaire. As explained in another book (Nuttin, 1984, Chapter 5), such goal objects and behavioral projects are concretizations or canalizations of needs as developed in specific situations. They are assumed to occupy virtually the individual's mind and to have an impact on his behavior: The subject *tries* or *intends* to realize them, he *hopes* that they will happen to him or that he can attain or avoid them; he *works for* their realization, etc. They are objects he is concerned with, or which interest him in one way or another.

The Motivational Induction Method is not only intended to study time perspective, but also to analyze the *content* of motivational aspirations of different groups of subjects in different experimental or clinical conditions. Empirical studies exploring the whole range of subjects' conscious motivations and aspirations are rather rare (1). Psychologists, as far as they are interested in motivational contents, generally limit themselves to an interpretative analysis of clinical experiences and personal intuitions when they develop systematic lists of human needs and motives. Research is usually done on one well-defined type of motivation, such as the need for achievement, affiliation, power, altruism, etc.

Unlike these studies, the Motivational Induction Method (MIM) tries to create a situation that is optimal for the subject to express spontaneously a large variety and range of personal motivational goals. In this, we agree with Murray's (1958) thesis "I am in favor of exploring the conscious motivations revealed only by direct methods, (...) I am in favor of direct methods (p. 185)". Research requires, however, that the motivational goals be expressed in a more or less standardized way, so that the results can be compared and analyzed in terms of broader motivational categories (see also Murray et al., 1938; Stein, 1947; Santostefano, 1970; for a comparison of T.A.T. and MIM, cf. Lens, 1974).

I. THE INSTRUMENT

The instrument includes two small booklets with 40 and 20 pages respectively. Two shorter forms with 30 and 15, or 20 and 10 pages also exist (see Appendix A). On top of each page, a sentence beginning (a *motivational inducer*) is printed or typed. All sentence beginnings are formulated in the first person singular, so that the subjects apply them spontaneously to themselves (for example: *I try to. ...*). The verb always expresses a tendency, an effort, desire, intention, etc. The subjects complete each sentence and, by doing so, they express an object of personal motivation. The sentence beginnings in the first booklet are so formulated that they induce *positive* motivational objects. The second one asks for negative objects, objects that are avoided, feared, etc.

Together with the booklets, the subjects receive an instruction sheet to follow when the instructor reads the instructions aloud. The confidential character of the study, its anonimity, and its purely scientific goal are stressed. The instructions can be found in *Appendix A*. It is also explained - especially with high school pupils as subjects - that it is not important to write grammatically correct sentences, but to express whatever object of *personal* tendencies, efforts or desires comes to mind when reading the inducers.

It is advisable to ask the subjects to consider not only the objects that they are presently most concerned with, but to consider all the goals that they actually pursue and all the objects that they are actively interested in. Thus, the MIM is not a free association technique, although subjects are encouraged to react *spontaneously*, i.e. without selecting or censoring their answers for social undesirability. At the same time, an atmosphere of privacy is created. Subjects are seated at a certain distance from each other and nobody with an authority relation to the subjects (e.g. teacher) is present.

The list of 40 positive and 20 negative inducers, as well as the two shorter versions and an instruction sheet, are given in Appendix A.

Construction and structure of the MIM

After many preliminary studies, the motivational inducers were selected from a long list of verbs and verbal expressions indicating different modalities of motivation. It must be clearly understood that our purpose in using a rather long list of inducers is to obtain a great variety of motivational objects, without making any suggestion about possible areas of motivation. In fact, the high number of inducers increases the probability that each subject will touch upon a more or less broad spectrum of his motivations, instead of concentrating on one or two topics. At the same time, it becomes possible to express different objects belonging to the same motive, and even to repeat one or more motivational objects. In some cases, it may be desirable to ask the subjects explicitly to consider all the areas of their activities, interests, and personal motivations, instead of limiting themselves to one or two fields of interest (perseveration tendency). With that view in mind, a complementary technique - the so-called *Motivational Inventory* - has been worked out in which the subjects are presented with a list of possible motivational areas (see *Appendix B*).

A negative aspect of our long list of inducers is that subjects have the impression that several inducers are repeated several times. In order to prevent their objections against it, the instructions explain that this is done intentionally in order to give them the opportunity to express as many objects as possible - but only one on each page - in all fields of their personal motivations and concerns. It is added, however, that they are allowed to go on to the next page when a personal motivational object does not come up in their mind. In fact, they are explicitly discouraged from expressing impersonal or inauthentic goal objects. However, when the same goal object, or a similar one, comes up in their mind, they may repeat that object even though they already expressed it on a previous page.

An advantage of our long list of 40 positive inducers is that it allows us to introduce the variety of motivational modalities with different degrees of activity, intensity, and realism. The following categories are represented. The first category of sentence beginnings expresses a general tendency for motivational objects through verbs such as: *to desire, to wish, to hope, to be inclined to*, etc. The same general tendency is present in inducers referring to *satisfaction* or *reward* (Nrs 4, 18, 25). The second type of inducers expresses activities such as planning or decision making: *I am resolved to* (Nr 24), *I am determined to* (Nr 9), etc. Finally, several inducers express a certain ongoing activity or effort: *I am working towards* (Nr 2); *I try to* (Nr 7); *I am striving (to*, or *for)* (Nr 16), etc.

Within each of these categories, additional modalities can be distinguished:

- Degree of intensity: *I have a great longing (for*,or *to)* (Nr 17); *I definitely have the intention to* (Nr 13); *I will do everything possible to* (nr 28).
- Allusion is made to possible obstacles: *I would not hesitate to* (Nr 15); *At all costs, I am willing to* (Nr 36).
- Some inducers express a certain hesitation, or are formulated in the conditional tense: *I would like so much* (Nr 21). We found that this type of inducers facilitates the expression of less desirable wishes, or motivations for which the subject is less certain or less decided about (see also the list of negative inducers).
- Inducers expressing a strong intention, a decision, or an ongoing effort have a higher level of realism than inducers with only a general tendency such as *I long for*, or verbs in the conditional tense such as *I would like*. To obtain goals or aspirations that the subject himself considers as less real, we selected the inducer *I am dreaming of* (Nr 12), with the specification that we do not ask for night dreams.
- The inducer 'As soon as possible, I would like...' (Nr 39) mentions a certain temporal urgency. This sentence beginning suggests a certain future time perspective (the immediate future), that has been avoided in the other inducers. For the same reason, the number of verbs in the future tense has been limited to five.

List of negative inducers(2)

Preliminary applications of the MIM showed the advantage of having all the negative inducers in a separate booklet. This is usually completed after the first booklet with positive inducers. It was found, in individual applications, that some adults are reluctant to express the objects of their fears in a straightforward way. Therefore, some negative inducers are formulated in the conditional tense, for example: *I would oppose it if* (Nr n3); *I would not in any way like it if* (Nr n8). In addition, the conditional tense allows for the expression of objects that are actually not immediately threatening.

There are only 20 negative inducers (15 or 10 in the shortened versions) because, generally speaking, subjects formulate much fewer negative than positive motivational objects. Moreover, some subjects do formulate their responses to positive inducers in a negative way: *I hope not to fail* (rather than *I hope to succeed*). This tendency is worth to be studied for itself. In fact, some theories claim that the tendency to avoid displeasure (or negative goals in general) is more fundamental than the tendency to reach positive goals (pleasure).

The negative inducers can be classified in the following categories:
- The sentence beginnings that induce a general negative attitude or tendency with different degrees of intensity: *I don't want* (Nr n2); *I would not in any way like it if* (Nr n8).
- The inducers that express a negative *affective* reaction towards an object. Sometimes the object is assumed to exist already: *I think it is sad that* (Nr n4). We avoided inducers expressing very specific affective or emotional reactions, such as anger, shame, etc. The verbs are usually in the conditional tense: *It would displease me very much if...* (Nr n1). In other cases the object is threatening or may become threatening. *I am afraid that...* (Nr n9); *I fear that...* (Nr n19).
- The inducers that refer to a negative *behavioral* action (*to avoid, to try to avoid, to oppose*): *I would oppose it if* (Nr n3); *I try to avoid* (Nr n6). One inducer asks for an object the individual avoids to *think* about: *I don't like to think that...* (Nr n14).

N.B. *The Motivational Inventory* that was mentioned as a subsidiary technique, is described in the Appendix B.

II. MOTIVATIONAL CONTENT ANALYSIS

MIM sentence completions yield an abundance of motivational objects, i.e. the content of individual tendencies. This content can be analyzed and investigated with relation to experimental and differential variables in research on motivational tendencies. The general motivational tendency or need that is at the origin of a specific object can usually be identified. To that effect, a 'Content Code' has been developed and is described in the *Manual of Motivational Content Analysis* (cf. Chapt. 7). In the present section, we limit ourselves to a description of the main motivational components that were discovered in our analysis.

The purpose of motivational content analysis

Independently of their temporal localization, which will be investigated in the next chapter, MIM sentence completions can be classified in different categories on the basis of their content elements or components. For example, a father's motivation to better understand his children is related to a social 'object' (his children), and the activity involved in the relationship is of a cognitive type (to understand) (3). Elsewhere, we define motivations in terms of the *categories of objects* involved and the types of the subject's *interaction* (activity) with these objects (cf. Nuttin, 1984, Chapt. 2 & 4). Hence our purpose is to analyze motivations in terms of these two *components*. While the psychology of human needs is mostly limited to a rather abstract classification, we think that an analysis of the components, and their various combinations, will lead to a better understanding of the internal structure and composition of motivations. This approach follows logically from our relational theory of motivation, in which motivations are defined in terms of types of required relations with specific categories of objects (*ibid.*).

A preliminary remark must be made. At first sight, the abundance and variety of the content of MIM-sentence completions make it very difficult to classify them in a meaningful way. However, it appears that this great variety in verbally expressed motivational data is an interesting phenomenon in itself. It reflects the flexibility of human contacts with an infinitely diversified world. The possibilities of our perceptual and behavioral interactions increase as a function of the complexity of the human psychophysiological organism. The versatility of verbal expression adds to this richness and variety of behavioral functioning. A subject who says that he tries to become more generous may be expressing the same thing as the subject who says that he wants to help his friend financially. However, the first person is talking about acquiring a personality trait (generosity), while the second is expressing himself in terms of a concrete interpersonal relationship. The study of the multiplicity and variety of verbal expressions about behavior and its motivation can be compared with generative and transformational grammar as studied by linguists. Nowakowska's (1973) publication on the language of motivation is also related to this problem although in a very different context. This problem seems to be typical for all types of content analysis.

This complexity of verbal data expresses the flexibility of structures and processes in human thinking. It is a positive aspect, but it is difficult to handle. With regard to our problem, the variable that brings unity to this verbal and behavioral complexity is the subject's final goal or anticipated outcome of action. Indeed, the goal, or the intrinsic object of an activity, defines the nature of a behavioral act and its motivation (Nuttin, 1983). The behavioral complexity is most evident, however, in the variety of intervening means-structures leading to the goal.

The main categories of motivational objects and interactions

Our content analysis of motivational *objects* as collected by the MIM will be facilitated by a first classification into four main categories with motivational relevance. The four types of objects involve different forms of interaction or relations between an individual and his world.

There is, first of all, the subject himself who is for himself a very specific object of interaction that we may call the *Self*. It includes all aspects of the individual as he is perceived by himself and as he appears to others with his physical, psychological, and social characteristics. It is found that very many MIM responses refer to different aspects of the self.

The second class of objects consists of *'other people'*. They have the same basic characteristics and functions as the subject himself (*alter ego*). These common functions create *reciprocity*, so that these objects also act as subjects who know and perceive, love or detest each other. Reciprocity is at the origin of specifically social motivation: One is interested in the other's feelings about oneself. The *kind of relationship* that exists between the subject and another person has a strong effect on the motivational meaning of that person for a given subject (for instance, one's husband, friend, girl friend, or people in general). Different motivations are associated with these different complementary aspects of human objects. For example, the affectional attachment to one's wife is qualitatively different from the affection for one's father or child. But all these relations have a common and unique characteristic: they are interpersonal and social (4).

The third, and very heterogeneous category is composed of all *natural* as well as *man-made* material objects. They derive their specific motivational meaning from the role they have in human behavior. We try to manipulate them or transform them. Our behavioral relations to them are different from those we have with people (second category). An important subcategory within the group of material objects are those that are 'required' for satisfying biological needs; for example: food-objects and objects that give protection and shelter. Human subjects act on these natural objects in order to make them more adapted to their needs and specific life circumstances (man-made objects).

The fourth class of behavioral 'objects' is of a very special nature: It consists of conceptual realities, i.e. objects of ideation and products of man's cognitive activities. We refer here to constructs such as science, ideology, philosophy, religion, political and social institutions, values such as freedom, truth, objectivity, reality, justice, independence (Nuttin, 1984, Chapt.4). Notwithstanding their intangible nature, these conceptual objects have often a very strong motivational force. They may become more important than material belongings and objects of personal interest. We do not consider them here as a world of realities in the sense of Popper's (1975) or Eccles' (1970) third world, but as simple motivational objects that have a real impact on behavior. In our *Content Code* of the MIM we call them "transcendental" objects. They are not very frequently expressed in MIM sentence completions. Transcendental objects seem to be strongly associated with the need for selfrealization, for autonomy, and for cognitive contact with the universe as a whole.

Besides these four categories of objects, several types of *activities* or behavioral *relations* with objects are to be distinguished in our content analysis. Some of these activities are typical for one category of objects only. For example, people are motivated to communicate and to exchange feelings and ideas with the second category of behavioral objects only, i.e. with other human beings: They are motivated to enter into behavioral contact with them (social or interpersonal relations). Two other types of social interaction with people are expressed in our MIM material. Some subjects expect others to take the initiative of entering into relation with them. The type of contact wanted takes three different forms: (a) The subject wants a certain type of affection, such as sympathy, friendship, love, forgiveness, etc.; (b) He expects some positive evaluation, such as appreciation, estimation, respect, regard, approval, etc.; (c) Finally, he is interested in receiving some form of help, support, assistance, cooperation, direction, encouragement, etc. The last type of social interaction is altruistic in nature: The subject wants certain things to happen to someone or tries to avoid the occurrence of some negative event, all for the benefit of others. These altruistic motivations are often formulated for a certain group of people (e.g. the poor; the Indians); so we call them humanitarian motivations. The opposite of such positive tendencies are aggressive interactions in which an individual is motivated to harm others and wants something negative to happen to them.

Besides social interactions and contacts that are limited to one category of objects, *viz.* to other human subjects, some behavioral activities may be directed towards the four classes of objects mentioned. Cognitive activities are the most important in this category. Thus, an individual may be motivated to know *himself*, to understand his *friends*, to be able to explain *natural phenomena* and *technological achievements*, to find *truth* or to study *existential problems*.

Two other motivationally important categories of human activities are *work* and *play*. These activities are less characterized by specific types of interaction with objects than by different kinds of effect that they intend to bring about. The same activity - such as reading a book or doing manual labor - may be work for one person and play for someone else.

The last type of behavioral relations that subjects are motivated for, is to *possess* the objects they are getting in touch with . This relation is mostly - although not exclusively - limited to man-made objects.

The analysis just made is limited to classifying the main kinds of objects and the types of activity or relations that are implied in the motivations as expressed by large groups of subjects. Each of these elements is conceived of as a *component* of a specific motivational entity. As we will see later, these components may be combined in different ways and constitute a broad variety of specific motivations. Thus, a motivation to better know or understand oneself implies some concern with oneself and a cognitive or explorative tendency, whereas the motivation to understand one's friends combines social and cognitive components (see below, p. 154).

The motivational components just described in terms of objects and behavioral relations are translated in our Motivational Content Code into ten *main categories* and about one hundred *subcategories* of 'motivational objects' (cf. Chapt. 7). The number of subcategories can be increased or reduced as a function of the specific object of a research project.

Several specifications have been added to this analysis of motivational components. They generally express motivational *modalities* such as, for example, the presence of a *conflict* or an *obstacle*, the fact of being *satisfied* with an actual situation or desiring *change* or *development*, a striving for *maximal* results etc. (cf. p. 159). Also the number of these modalities can be increased or decreased according to the goal of a study.

Each of these categories, components, and modalities has a specific symbol in our Content Code (a combination of letters and numbers). We refer the reader to the Manual for the details about the Motivational Content Analysis (cf. Chapt. 7). It may suffice to say here that the structure of the code in terms of various components enables one to identify easily all motivations that imply in one form or another such elements as cognitive activities (symbol: E), social interaction (C), professional activities (R_2), the individual Self (S), specific physical or psychological aspects of the subject's personality (Sph or Sps), recreation and leisure (L), possessions (P), etc. From the viewpoint of research, it may be interesting to investigate the changes within the components of the subjects' motivations as a function of some experimental or differential variables (see the *Conclusion* of Chapt. 7). As to research in the field of time perspective, it should be noted that motivational content analysis is also important, since the temporal localization of a motivational object or plan is usually related to its content.

After this brief review of the motivational components as

revealed in the MIM-sentence completions, a discussion on the validity and reliability of the MIM is in order. Even though this method is not intended to be an *individual* diagnostic instrument, we still have to know to what extent and with what degree of reliability the MIM does provide us with representative samples of the motivational objects and concerns of specific *groups* of subjects.

III. VALIDITY OF THE MIM

Before approaching the question of validity, we should first define what we expect from the MIM, what its possibilities and limits are. The MIM is neither a motivational test, nor a technique for measuring the strength of different motivational tendencies. It is intended to obtain a representative sample of the motivational objects of a group of subjects. The first question, then, is the extent to which the MIM-sentence completions are able to reveal a sample of the real motivational objects that activate and direct subjects' overt and covert behavior.

Face validity of the MIM

Before evaluating the validity of an individual's verbal expressions of his own motivations, an important distinction must be made. Do we want the MIM to give us some concrete motivational objects of an individual, or rather to reveal the more fundamental motivational tendencies underlying those specific goals? Let us take the example of a subject with a MIM-sentence completion saying that he or she intends to become a medical doctor. Assuming that the subject fully collaborates, such a communication is taken as truthful information that: 1. The expressed goal is an object that the subject consciously aspires or strives for; 2. That the motivational object expressed has an activating and guiding effect on different overt and covert actions of the subject.

On the other hand, we know that the subject is not always aware of some tendencies or variables that, perhaps unconsciously, codetermine his goalsetting. His own explanations of his intentions may be unfounded rationalizations. Generally speaking, it may be very difficult for a subject to detect the mental processes underlying a specific motivational goal or a certain 'causal' attribution of his behavior. The conditions in which such subjective attributions can be valid should be specified (Nisbett & Wilson, 1977; Smith & Miller, 1978).

The MIM, however, only assumes that people are able to know and communicate a great number of the concrete goal objects that they pursue in their daily life (to become a medical doctor; to be financially independent from my parents, etc.), and that most of the time these goal objects have some motivating effect on their behavior. We agree with Allport (1961) and others that human behavior is largely regulated by man's conscious goals and behavioral projects: "We live our lives, at least in part, according to our conscious interests, values, plans, and intentions" (p. 216). The importance of unconscious processes in the development of goal setting is not denied and some unconscious cravings may underly conscious goals and plans. Most of the time, one does not know the processes underlying one's concrete desires. This is even true for hunger and thirst; but when a subject is motivated to get food, he is conscious of this goal, and it is this conscious goal that directs his behavior, although the origin of his hunger may be unknown to him.

Moreover, it happens in social life that one's real intentions and goals are disguised behind verbal expressions that mean almost the opposite of what is actually said. This is not only true in formal diplomatic language. This may explain why the relationship between what is said and what is done is very weak in certain social circumstances (Argyris & Schön, 1974; 1978). However, this does not deny a person's possibility to express his concrete and authentic motivational goals, at least when some favorable social and personal circumstances are created for doing so. The MIM should, therefore, be administered in conditions that allow the subject to express directly and anonymously the objects of his personal and even intimate aspirations and intentions.

The second preliminary question has to do with *social desirability*. It can not be denied that, even in favorable conditions, social desirability has an influence on the communication of motivational objects, as some studies have shown. Two remarks should be made in this regard.

First, it may be true that certain inhibitions prevent a subject from saying, for example, that he wishes his rival's bankruptcy, or that he would like to do harm to his ennemy. But the same inhibitions that prevent him from saying it may also prevent him from doing it. In other words, such motivations often have more impact on a subject's fantasies and wishes than on his actions. After all, it may be desirable that motivational psychology remain interested in motivations that produce overt behavior and not be confined to those underlying dreams and fantasies! In addition, one should not forget that social and personal inhibitions and censures are also based on fundamental human motivations.

Second, it is reasonable to assume that the motivational category to which an inhibited desire belongs (for example aggressive competition) will be expressed in other specific motivational goals for which there is little or no inhibition and that are more directly related to the individual's overt behavior (such as professional competition and promotion). Moreover, it appears from our content analysis of MIM responses, combined with clinical interviews, that subjects are not only eager to express anonymously their true motivations and concerns (Murthy, 1963, p. 54; Meyer, 1966), but also that they do not hesitate to communicate confidential, intimate, and sometimes even compromising motivational objects (Goethals, 1967; Lefebre, 1969; Moors, 1972). Noncooperative subjects and inauthentic protocols can usually be discovered easily and eliminated. However, it is probably true that some types of motivational goals are underrepresented in MIM responses. There are some categories of objects that are not readily placed in a social context, such as transcendental concerns and aspirations (cf. p. 156). Further, objects that are taboo in some cultures are underrepresented in some groups. Vanden Auweele (1973) found that, for adolescents, in a sexually tolerant society, 25 out of their 155 responses in the field of sensorial pleasure were explicitly sexual desires. For a comparable group in a much less tolerant society, the proportion was only 2 out of 203.

Another general difficulty in the exploration of motives resides in the fact that satisfied needs, or needs for which the satisfaction poses no problem, are seldom expressed. Subjects are often not even conscious of the fact that these needs exist. Needs only manifest themselves when their satisfaction becomes problematic.

Finally, a few words about the relation between aspirations or goals on the one hand, and actions on the other. To what extent do motivational objects that are expressed in MIM-sentence completions have an impact on the subjects' activities? In fact, MIM responses may refer to desires, wishes, aspirations, and goals without giving any information about a subject's overt behavior to attain or avoid these positive or negative motivational objects.

The following two considerations are related to this problem. First, several MIM-sentence beginnings imply an activity in their formulation; for example: *I am working towards..., I am striving (to,* or *for)...; I am preparing myself (to,* or *for)..., I definitely have the intention to...*. The responses induced by such sentence beginnings should, normally, include goal objects that are being actively pursued (*face validity*). On the other hand, motivational objects often have a behavioral impact outside the field of overt behavior. As noted above, the influence of conscious motivational objects on affective and cognitive aspects of behavior and on human life in general may be of the utmost importance for the subject. Therefore, the impact on overt behavior should not be our only criterion.

In summary: The MIM is intended to sample motivational states that are cognitively processed into goal objects, means, aspirations, and desires. It does not uncover the underlying tendencies nor the processes that made the individual formulate those specific motivational objects. Our method assumes that goals, projects, or intentions have some impact on overt and/or covert behavior. It further assumes that, in favorable conditions, such desires, projects and intentions can be verbally expressed within the limits mentioned. MIM responses, therefore, have face validity.

Validity studies

If the MIM is a valid instrument to obtain a sample of the motivations of a group of subjects, it should reflect the effect of an experimental manipulation that stimulates a certain motivational category (for example, achievement motivation for academic performances). Cossey (1974) created two different degrees of this specific achievement motivation by using a method borrowed from McClelland et al.(1953). A group of 34 female and 27 male undergraduates were presented with a test that was introduced as highly predictive for academic results. These instructions were intended to arouse strong achievement motivation. By using arbitrarily high norms, all the subjects failed to reach the criterion, so that the aroused motivation was not reduced by success experiences (*failure condition*). The same test was administered to an equivalent control group of 32 female and 24 male undergraduates for purposes of initial item-analysis (*neutral condition*). Nothing was done to stimulate the achievement motivation. Immediately following this test, the MIM was administered by another research fellow to both groups and presented as part of a totally different research project.

The results confirmed the hypothesis: 1. The male subjects in the failure condition expressed a significantly higher relative number of positive goal objects related to study behavior than the male subjects in the neutral conditions (22.12 percent and 15.10 percent; $p < .002$); 2. The relative frequency of the other motivational categories did not differ between the two groups: They were not affected by the experimental manipulation (5). These results show that the MIM is sensitive to experimentally induced motivational differences between groups of male students.

Craeynest (1967) found the MIM to be sensitive to situational effects on another type of motivation. The MIM was administered to a group of undergraduate students at the end of a long morning (between 12 a.m. and 1 p.m.) and to an equivalent group shortly after lunch time. In their MIM responses the first group referred significantly more to food and to eating behavior than the second group did. A study in Tanzania showed that motivational goals related to food, shelter, etc. were the most important category (25 percent of the responses) for a group of poor people. This same category ranked only seventh (4 percent) in a group of more well-to-do people within the same culture (Meyer & Grommen, 1975).

Lens (1974) found that soldiers who were hospitalized in a military mental clinic because of neurotic disorders expressed significantly more motivational concerns about their psychological health than an equivalent group of sick or injured soldiers in a normal military clinic did about their physical health. It seems reasonable to assume that subjects in the latter group were less concerned about their health than the former group of neurotic subjects for whom the gravity and duration of their state of health was much less clear.

Other studies (Goethals, 1967; and Lefebre, 1969, with prisoners; Noterdaeme, 1965, with college graduates in compulsory military service and an equivalent group starting a professional career) showed that the content of MIM-sentence completions reflects very well the subjects' unfulfilled desires and aspirations.

Goethals (1967) found that a group of 30 male prisoners (condemned for murder) expressed much more frequently desires for consideration, affection, and help from others, than did an equivalent group from the population in general. This study showed also that only 8.6 percent of the goals expressed by the prisoners were situated within their detention period, while 65.6 percent referred to subsequent time.

These studies allow us to conclude that the MIM can give a sample of the motivational concerns of groups of subjects.

The problem of the *stability* of the motivational contents expressed in the MIM will be treated in the next section on reliability.

IV. MIM RELIABILITY AND STABILITY OF MOTIVATION

The MIM can only be considered as a reliable method of collecting samples of motivational goals if its results have a sufficiently high degree of stability. It is important to realize that the reliability problem is more complex for methods such as the MIM than for intelligence tests and even most personality tests. Intelligence and most personality traits are assumed to have a high degree of stability, so the lack of stability in test results is attributed to the unreliability of the test itself. However, for motivational objects, the degree of stability with regard to the specific objects that are actually desired and feared is a problem in itself. We know that the specific motivational objects pursued may change as a function of situational circumstances, but we also know that it is very difficult to change one's general motivational orientations, such as one's scale of values or interests. In this sense, motivation is a stable element at least in its main orientations and structures. The same type of opposition can be found between the concrete motivational objects pursued by an individual and the limited number of fundamental human needs.

There are two reasons why we may expect a certain stability within MIM-data.

First, because of the large number of inducers, we may expect that the expressed motivational objects reflect not only the subjects' concerns at the very moment of testing, but also a broader scale of virtually present objects in their current life situation. The latter will manifest more stability. Moreover, and what is most important, we do not register when analyzing the MIM data, the concrete motivational objects, but rather the main motivational categories and subcategories in which the motivational objects are to be classified on the basis of their motivational meaning or content (see our Motivational Content Analysis). These motivational categories correspond more or less to general motivational orientations and, as such, will be more stable than the concrete objects, which are much more dependent on circumstances.

However, it may happen that a subject, instead of expressing motivational objects from all fields of his activity, will focus his attention on one subfield and stay with it. This results, of course, in repetitions (perseverations) and in differences between two consecutive applications of the MIM, since the subfield considered may be different in the two cases (cf. the *Manual*).

In conclusion, a certain instability in MIM data may be caused by the instability of the motivational objects themselves. However, the very fact that in processing our data, the more general categories of motivation are taken into account - and not the concrete motivational objects as such - gives the data a higher degree of stability.

Before presenting some data on the reliability of the MIM, we should first discuss the reliability of our coding technique.

Coding reliability

To check the coding reliability, the analysis and coding of motivational contents is usually done by two trained judges working independently; in some cases, the coding has been repeated by the same judge after a certain time interval. The motivational categories in which the data are classified are mutually exclusive.

The coding reliability is measured by calculating the relative number of corresponding codings in the two analyses. Trained judges reach an intercoder-reliability of 90 percent to 95 percent for responses to positive inducers (Lens, 1971; Verstraeten, 1974; Cossey, 1974; Nuttin and Grommen, 1975). For less trained judges the correspondence usually reaches 80 percent to 90 percent of the responses. Coding reliability is also somewhat lower (88 percent) for responses to negative inducers. After discussion of the non-corresponding codings, the correspondence may increase to 100 percent (Cossey, 1967; Craeynest, 1967) for the positive inducers and to about 93 percent for the negative inducers (Lefebre, 1969). Goethals

(1967) and Cossey (1974) recoded their MIM-data after 8 and 18 months, and disagreed with their first coding in about 10 percent of the responses. Almost all the inconsistencies could be solved by re-examining the responses in collaboration with a senior coder. A few answers had to be considered as "unclassifiable". Bouffard (1980) recoded a sample of his MIM data after a three-month interval. For 93 percent of the responses, the two codings were identical. He reports intercoder-reliabilities of 90 percent and 87 percent. When he applied Cohen's (1960) formula to take into account the degree of correspondence between two coders that may be caused by chance, he obtained a K-coefficient of 0.856. This value approaches the maximal possible value of K for his data, namely 0.93.

For the sake of comparison, it can be mentioned that T.A.T. coding reliability varies between 80 percent and 90 percent in some studies, and between 92 percent and 95 percent in other studies.

Stability of MIM-data

To what extent do two applications of the MIM with the same group of subjects, or with two equivalent groups, produce the same motivational configuration?

Craeynest (1967, p. 37-38) readministered the MIM after a two-week interval. A rank correlation of .95 was calculated between the main motivational categories. In only 7 out of 95 motivational subcategories was the frequency significantly different ($p < .05$). Some of these differences could easily be explained on the basis of situational differences between the two applications (6).

Craeynest (1967, p. 114) divided a group of 70 subjects into two aselect subgroups. All the subjects were tested at the same time in one large room. The rank correlation for the main motivational categories was .96. The difference between the two subgroups was statistically significant (2.06 percent and 3.28 percent; $p < .05$) for only one main category (with a low frequency). Verstraeten (1974) did the same for a group of 118 adolescents and found a correlation of .97 for the main motivational categories. For the sake of illustration, the relative frequency of the 10 motivational categories in both subgroups is given in Table 1.

Cossey (1974, cf. supra p. 57) compared the frequency of the motivational categories in his two groups, except for the category

TABLE 1

Percentages of motivational objects in each of the ten main categories for two equivalent groups of female adolescent subjects (Verstraeten, 1974)

Cat.[1]	S	SR	R	C	C_2	C_3	E	T	P	L
Gr. A	20.34	4.54	7.12	22.10	8.07	9.36	4.88	0.47	11.46	11.66
Gr. B	18.76	5.83	7.92	21.74	7.92	10.08	4.63	0.95	9.70	12.48

[1] For the meaning of the initials, see Chapter 7

that was - as hypothesized - affected by the experimental manipulation (achievement motivation in study behavior). Using the KMS (Siegel, 1956), he found no significant differences between the two groups, neither in the main categories, nor in his 50 subcategories.

Moors (1972) used a special kind of split-halfreliability test. He administered the MIM in its usual form to a group of 196 students (college level). He divided his subjects into two aselect subgroups (Subgroups A and B), and the MIM sentence completions into two halves: The responses to the even-numbered inducers were classified in one half (Responses I), and the responses to the uneven numbered inducers in the other (Responses II). He then grouped the Responses I of Subgroup A with the Responses II of Subgroup B in one group, and the Responses II of Subgroup A with the Responses I of Subgroup B in another group. By doing so, he erected two groups of MIM sentence completions, composed as follows:

Group I: Responses I of Subgroup A and Responses II of an equivalent Subgroup B

Group II: Responses II of Subgroup A and Responses I of Subgroup B.

This technique is intended to cancel out all possible differences not only between the two subgroups of subjects, but also those that may exist between the two halves when the total list of inducers is divided into two parts (for example even versus uneven numbered inducers) as it is usually done to estimate the split-half reliability. Moors did not find any significant differences in the frequency of motivational main- and subcategories between the two groups of MIM responses. As to the perseveration tendency, which could have influenced this comparison, Moors found that only in 1.5 percent of the cases, were sequences of two similar motivations (perseveration) given; sequences of three similar responses were found in only 0.5 percent of the cases.

As a conclusion: The results of these studies show that the MIM, when applied in optimal circumstances and analyzed by trained coders, has a sufficiently high validity and reliability to sample the motivational objects of groups of subjects.

V. ADDITIONAL POSSIBILITIES OF DATA-COLLECTING BY MEANS OF THE MIM

Some minor changes of the MIM-instrument make possible the study of other aspects of time perspective and mental content. The inducers can be formulated in such a way that the subjects will express their motivational objects and aspirations inasfar as they are members of a certain social group, or inasfar as they function in one or other social role.

For example: We, women, we hope...
As a housewife, I strive to...
We, French speaking Canadians, we want...
As a Jew, it displeases me that...

The qualification suggested in the examples given, and many others, can be added to each positive and negative inducer.

A study of this type was done with Huron Indians as subjects. First, the MIM was administered in its usual form. In the second application, the sentence beginnings were formulated in such a way that the subjects were asked to answer as Hurons: *As a Huron-Indian, I...*; or *We Hurons....* The two sets of data permit the study of conflicts that may exist between individual motivations and social roles. This technique also allows for the study of thematically different time perspectives (cf. p. 25). The future time perspective of young women 'as housewives' may be different from their future time perspective as far as they are 'careerwomen' or members of a certain club, etc. This approach has been fruitful in anthropological research.

Another adaptation of the MIM makes it suitable for the study of non-motivational mental contents, such as objects of cognition and emotion. By replacing the motivational verbs in the inducers by verbs expressing some purely cognitive or affective activities, the content of thoughts and feelings can be sampled. It is obvious, however, that, with such sentence beginnings, the method can no longer be labeled a *"motivation induction method"*. And as regards the study of time perspective, "thought objects" and even emotional objects are better suited to study the subjects' *past* time dimension in comparison with their *future* time perspective. Such comparisons between subjects' future and past time perspectives are important with regard to the study of preferential time *orientation*: Are these subjects more oriented towards the past then the present and the future? Motivational inducers direct the subject by definition towards the future.

Frequency of expression and subjective intensity of motivation

As mentioned earlier (cf. p. 57), Cossey (1974) found that an experimental manipulation of the subjects' strength of achievement motivation significantly increases the frequency of that motivational category in their MIM-sentence completions. One may ask the question to what extent the *frequency* of expression of a given motivational category correlates with the *intensity* of that motive in a given subject. In several studies the answer to that question is assumed to be positive, although few systematic investigations have been done to this effect. We limit ourselves here to data collected with our Motivational Induction Method.

After responding to the MIM, subjects were asked to indicate, on a three-point scale, the strength or intensity of each expressed motivation (weak - strong - very strong) (7). In an initial analysis, Cossey (1974) rankordered the motivational categories for their frequency and for the mean subjective intensity of their motivational objects. This was done for both the positive and the negative inducers separately. The rankcorrelations for the main categories and for the subcategories of motivations were all statistically significant with values ranging from .36 to .74. However, no significant rank-correlation was found when - for each category - the subjects were rankordered for the number of motivational objects expressed in that category, on the one hand, and for the mean subjective intensity of their motivation in that same category, on the other. Cossey also found a significantly higher degree of subjective intensity for the experimentally manipulated achievement motivation in the failure-condition than in the neutral condition.

Duces (1968) too found significant correlations - although they were not high - between the frequency of motivations within main categories and the subjective intensity of those motivations. The category of transcendental motivations was an interesting exception. A group of subjects who expressed either no, or very few motivational objects in this category, maintained, in the second stage of the study, that their transcendental goals and aspirations were stronger than their other motivations. This result confirms that some types of motivational goals are not easily expressed in MIM sentence completions.

We may conclude that the most frequently expressed categories of objects are also the most intensively motivated ones, but only at the level of motivational categories, and not at the level of individual subjects (Cossey, 1974).

NOTES

(1) We refer to the content analysis of clinical interviews by Dollard and Auld (1959), the techniques used by Murray et al. (1948) and Stein (1947), to Combs' (1946a; 1946b) list of motivations, and to Santostefano (1970) and Pringle (1974) for the motives and needs of children.

(2) The numbers of the 20 negative inducers are preceded by the letter *n* (cf. *Manual*).

(3) We make here a distinction between the *object* (the children) and the *activity* or *action* related to that object (to understand). Very often, we use the concept *object* in a much more global meaning, as for example when we say that the object of

the housefather's motivation is *to understand his children*. The content analysis requires here to distinguish the 'material' object that is contacted (children) and the type of contact or activity involved, viz. understanding.

(4) To the extent that a reciprocal contact is possible with some animals, they form an intermediate category between inanimated objects and the *alter ego*.

(5) For the responses to the negative MIM-inducers, the difference was not significant. In both conditions, the female students expressed more study related goals than the male students.

(6) We remind the reader that for each application of the MIM the frequencies of the responses in the different categories are not independent of each other. An increase of frequency in one category for the second application will entail a decrease in the frequency of one or more other categories.

(7) In some of these studies the subjects were asked, after they completed the MIM, to indicate the degree of impact of each expressed goal on their behavior during the last weeks.

CHAPTER III

**MEASURING TIME PERSPECTIVE:
THE TEMPORAL CODE**

Positive and negative motivational objects as expressed by our subjects are situated in the psychological time dimension. They are events or situations that may occur, goals to be achieved, or things one intends to do. Some of these objects (1) may refer to a way of being and behaving that lasts for a more or less long period (e.g. to keep fit), or to qualities one wants to have from now on and for a more or less unlimited time (e.g. to be less impulsive). Other objects are less closely related to a period in the individual's personal life, but have to do with, for example, the development of his family, business, social class, or mankind in general (e.g. food supply in the Third World; unemployment, etc.).

Generally speaking, MIM-objects can be situated either in the past, the future, or the psychological present. But by their very nature, motivational objects refer to the near or distant future. Therefore, the analysis of the time dimension in motivational objects is about the *future*. This is why our method is not intended to study a subject's preferential or dominant orientation towards the past, present, or future (*time orientation*).

I. THE BASIC PRINCIPLE

Our temporal coding of MIM-objects is not based on the subject's personal estimation about the time in which the event may happen or the goal may be reached, but on the "normal" or "averarage" temporal localization of an object within the social group to which the subject belongs. For example, the time localization of an individual's motivational intention to marry is not coded in terms of his personal temporal planning, which we do neither know nor ask about, and about which he himself may be very uncertain, but in terms of the time period where, on the average, such an event occurs for members of the group to which the subject belongs. We will justify this so-called "objective" temporal localization in the next section.

The temporal coding consists in assigning to each object the time period in which it "normally" happens. The assignment is done by trained judges, specific coding rules having been developed in the course of preliminary research. Details about the coding technique and examples are given in the *Manual of Time Perspective Analysis* (see further in this volume).

'Objective' temporal localizations are, of course, approximative. The basic rule is to ask, for each motivational object, what is the time period an average subject implicitly has "in mind" when formulating that goal object. How far into the future does his "intentional look" go when he thinks about that object? In the same way as a visual look stops at an object that is situated at a certain spatial distance, an intentional look or cognitive-affective anticipation is directed towards an object that is expected to occur, or to be realized, in a more or less forseeable period. It is obvious, for example, that the individual who wants to smoke a cigarette thinks of a motivational object that, for people like him, must be situated in the very near future (in a few seconds or minutes). On the contrary, the behavioral project of building a house or buying a car refers to motivational objects that are situated in a more distant future for the young man who happens to be our subject. By their very nature, and according to what is usual in a certain group of people, most motivational objects have their "normal" place in time. *The totality of the temporal localizations of the motivational objects of a group of subjects gives an idea of the mean extension of their future time perspective.* The hypothesis on which the MIM is based assumes that the motivational objects of MIM-sentence completions occupy, at least implicitly, the subjects' mental life. As such, they also influence their global behavior. Thus, we can say that the temporal signs of these objects create the individual's active time perspective.

II. JUSTIFICATION OF THE "AVERAGE" LOCALIZATION

As explained in Chapter One, the *subjective* temporal localization of motivational objects is difficult and problematic. Before describing our technique any further, we should justify it. We call it an "average" or "objective" temporal localization, because it is not based on subjective estimations by the subject, but on the nature of the object and its normal or average occurrence in the social life of the group.

Many researchers in the field of time perspective limit themselves to vague and very broad localizations. They distinguish, for instance, two or three periods: the near, the distant, and the very distant future. But as soon as more precise localizations are required, difficult problems arise. Previous research has taught us that temporal localizations of motivational objects by the subjects themselves are very dubious and difficult to compare with each other. The subject himself often feels unable to do it. Without detailed instructions about the technique and the temporal categories to be used, subjects will apply very heterogeneous temporal units. Some use very vague temporal indications such as "it will happen much later", the meaning of which may be different for different subjects. Others answer with a number of years, but are rather unrealistic. Some motivational objects are very difficult, if not impossible, to localize in time (for example: *I would like so much to be completely independent of my parents; I will be very happy when my husband understands me better*).

Based on a good deal of preliminary work with a group of associates - in this context, the early work of Noterdaeme (1968) deserves a special mention - and on the basis of theoretical considerations, we developed our present technique. The theoretical principles can be summarized as follows.

The temporal signs that characterize the objects originate in the individual's experience of time as developed during the process of socialization (cf. p. 17). We have called this the social clock. The temporal localization that is learned in a social context continues to characterize an object, even when the subject intends to realize that object at a moment that is different from the social "norm". As noted earlier, the subject who plans to marry "very early" makes this subjective temporal localization with reference to the average age at which people get married in his cultural group or subgroup. When he thinks about marriage, he also 'has in mind' its temporal sign, i.e. its normal time localization. Our objective method considers, first of all, this normal or average localization, *which psychologically exists in the subject's mind*. The fact that this normal time localization can be assumed to exist - implicitly at least - in the subject's mind when he talks about his personal motivational objects is the basic justification for using that time period as the objective clue for our time perspective measurement. If a subject, however, explicitly mentions the future moment he has in mind, which rarely happens, we code this temporal localization, as for example when a student says he wants to get married "this year". As a rule, all explicit temporal information spontaneously given by the subject himself is taken into account when coding the time perspective.

The main practical advantage of our objective coding method for research purposes is that it is based on a stable system of psychologically justified rules established over years of application; these rules can be applied in a uniform way by trained judges in such a way that the results obtained are comparable. This is not the case when estimations are made by the subjects themselves on the basis of very different criteria and in heterogeneous terms.

However, the principle of objective coding should not prevent us from studying, if necessary for specific research purposes, the subjective time experiences and estimations of the subjects themselves. The release of a prisoner with a 20 year sentence can be objectively situated in time when we know the rules and practice for probation in his country. The degree of temporal distortion in the prisoner's mind when he expects to be free next Christmas can be evaluated with reference to the objective time distance. In such cases, subjective time localization may be nothing else than wishful thinking; the objective temporal code may serve as a reference mark to measure the distortion. In some categories of subjects, a distorted perception of future events and goals may be an important feature to be studied in its own right. Moreover, it is recognized that in a clinical study of personality, the subjective localization of that person's goal objects should be known in order to account for his behavior. To that effect, a method for comparing subjects' personal estimations with our objective codings has been described and tentatively tested (Nuttin, 1980, p. 81). However, as said before, our main purpose here is to measure the extension of future time perspective in *groups* of subjects as a function of differential or experimental conditions.

III. THE LIFE PERIODS OF SUBJECTS

Measuring a subject's time perspective implies two points of reference: The time period in which a subject finds himself at the present moment (time of subject) and the time period in which the subject's motivational objects are to be localized (time of objects). The temporal distance of objects localized in a certain life period (e.g. professional retirement) depends, of course, on the subject's age. A person who expresses negative feelings about his retirement anticipates an object that is very near in time if he is at the end of his regular professional career. But that same object is still far away for the college student thinking about his retirement. In our coding system, the time of objects is expressed either in absolute units such as days, weeks, or months from now, or in terms of socially or biologically defined life periods. The description of this *time of objects* is the main topic of our coding system. Before doing that, a few words have to be said about the time of subjects: How do we define and limit the life periods in which our subjects find themselves at the moment of participation in a study?

In our Western cultural context, a lifetime can easily be divided into three major sections: the preparatory or educational period, the productive or adult life, and old age (see Fig. 1). Certain adaptations, however, may be necessary for subjects from other cultures (cf. Bouffard, 1980, for Rwanda).

- *The preparatory or educational period* (symbol E) embraces:

    E_0: until schoolage (0-6);

    E_1: the primary school period (6-12);

    E_2: the secondary or high school period (12-18); also subjects between 12 and 18 years of age who do not attend secondary school are classified in this life period unless they are married;

    E_3: the period of post-high school education, including professional education (18-25).

- *The adult life* (code A) is assumed to begin in our culture when the individual begins to do productive work, becomes financially independent, or starts his own family. Because of its length, this period is divided into two or three parts:

    A_0: A transition period (18-25) which coincides with the E_3 period for those who go through a prolonged educati-

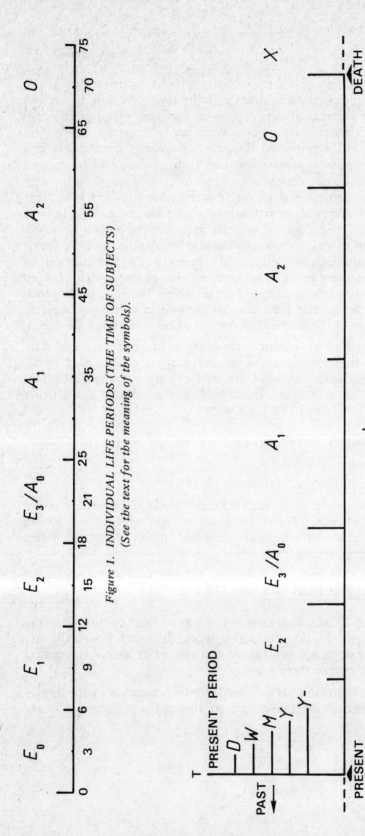

Figure 1. INDIVIDUAL LIFE PERIODS (THE TIME OF SUBJECTS)
(See the text for the meaning of the symbols).

Figure 2. THE TIME PERSPECTIVE SCALE (THE TIME OF GOAL OBJECTS)(*)
(See the text for the meaning of the symbols)

(*) This example applies for the motivational goals of 12 to 18 years old subjects. The immediate future (in calendar units) is situated in their present E2-period.

onal period;

A_1: from 25 to 45, first half of a professional career or adult life;

A_2: from 45 to 65, the second half of a professional career (until the age of retirement).

- *Old age* (code *O*): this final period starts at the age of retirement or at 65.

It is not surprising that the main periods within an individual's psychological life time correspond to the periods imposed by social and biological factors. In fact, each individual life develops within these contexts and social frameworks do have their psychological reality for most subjects (for example: the start of a professional career, retirement, marriage, the beginning and end of school education). It remains possible, however, that personally important events function as individual landmarks, establishing additional subdivisions.

IV. THE TEMPORAL CODE: THE TIME OF OBJECTS

The data to be coded are motivational objects expressed in MIM sentence completions. To understand the meaning of some objects and to situate them in time, it may be necessary to take into account the whole sentence and even the total context of all responses of the subject. A subject may formulate a response while implicitly (or explicitly) referring to, or thinking about, a previous one. It is therefore advisable first to read all the responses in a subject's booklet in order to have an idea of the totality of his motivations. Due to the context, the same response may be coded in different ways for different subjects. The context of a response, i.e. the other responses in the booklet, may change its meaning. For example, the response "*I wish to be the first*" has a different motivational meaning when it follows "I intend to run the New York marathon", or "I wish that many of my friends would take jobs in a developing Third World country". It was also found that the inducer, i.e. the sentence beginning, may affect the temporal situation of a response. It should be realized, however, that the diversity of inducers precisely intends to elicit different types of motivational objects.

The temporal code consists of a series of symbols that refer to the successive periods of the experienced time. It is not the result of an a priori conception, but of many empirical studies in close contact with our subjects' experience of the future and the past. It is based on the fact that most motivational objects are related to some life period. Goal objects related to study behavior are usually to be situated in one of the educational periods (E). Professional goals are localized in adult life (A). More daily activities such as eating, walking, going for a swim, seeing a friend, etc. are intended for the immediate or near future. Other objects such as buying a car or building a house take more time to be realized according to the situation of the subject: individual circumstances will codetermine if it is a question of months or years. In general, for a subject who recently started his professional career, the behavioral project of buying or building a house refers to the first half of his adult life (A_1). One *becomes* a medical doctor in a medical school (E_3), but one *is* a medical doctor, i.e. one exercices the profession, during one's entire adult life (A). It is, therefore, assumed that subjects, when talking about the objects mentioned, have their 'mental eye' looking at the corresponding time period (time perspective).

The temporal scale on which the motivational objects are situated is divided into two different parts. The first part concerns the rather *near future* and is subdivided into calendar units: the present moment, today, within a few days, this week, within one or two weeks, this month, next month, within a few months, this year, next year.

As soon as the distance is longer than one or two years, temporal localizations become more vague. They can no longer be expressed in terms of absolute distances from the present moment, but in terms of social periods or units: "before I finish college", "as soon as I will be married", "when I will have a job", etc.

The transition from the calendar scale to the scale in terms of the social clock, is characterized by an increasing lack of precision in subjects' direct experience of time distances. It is well known that an increase in temporal distance - in the same way as for spatial distances - makes the perception of the temporal localization less precise and even less "real". It is difficult to foresee, in a precise and clear way, objects that are situated 5 or 10 years from now.

It is in correspondence with these two ways of experiencing the future, that our temporal scale was constructed in terms of *calendar units* in the first part and in terms of *social clock units* in the second part (cf. Fig. 2). The detailed rules of temporal coding are explained in the *Manual of Time Perspective Analysis* (cf. Chapt. VI); only the general structure of our scale will be described here.

a. *The temporal scale in terms of calendar units*

> T *(test)*: the moment of the MIM application
> D *(day)*: within one or two days from now
> W *(week)*: within one week from now
> M *(month)*: within one month from now
> Y *(year)*: within one year from now
> $Y-$: within a period that may last somewhat longer than one year (within one or two years).

b. *The temporal scale in terms of social and biological units*

The symbols used for the second part of the temporal scale are the same as those for the life periods of the subjects, *viz.* E_1, E_2, E_3, A_0, A_1, A_2, O. They now refer to life periods in which the subjects' motivational objects can be situated. Additional symbols are used to make the time localization more precise, if possible. Sometimes it may be possible, for example, to make a distinction between goal objects situated at the beginning or at the end of a period (cf. *Manual*).

Two more symbols were added to this list:

L *(life)*: when the total future lifetime of the subject is explicitly mentioned as a global unit (for example: *I would like to live here for the rest of my life.*).

X : refers to objects localized in the subject's afterlife period, for example: a subject who is concerned with what will happen to his children after his death. This symbol is only used for objects referring to the personal lifetime of the subject. A small x is used for the historical future and for future social development in general (see *Manual*).

c. *The open-present (symbol: l)*

A rather large number of motivational objects refer to personality characteristics, abilities, or modalities of being.
For example: *I would like to be more friendly with people; I wish to be happy; I want to be understood*, etc.

As stated above, we ask: what time period do subjects have in mind when they think of such motivational objects? Our answer is as follows: they think about something they like to be or have right now, but that will go on in the future. For instance, a subject wants to be intelligent now and also later on. Although he mainly thinks about the present, his 'mental eye' is not enclosed in the present moment; it is directed towards a *future without precise limits*. The time perspective implied in this type of motivational objects is different from the time perspective that is limited to the present moment or the near future. The subject who writes *"I want to finish this test"* or *"I want to smoke a cigarette"* is thinking about the present moment. The subject who says *"I want to be happy"* thinks not only about the present, but also vaguely about his whole future life. He considers a much larger time period. We therefore classify this type of response in a time category labeled: the *open present*, that is, open to the future. Because it is difficult to estimate the temporal extension of such motivational goals, we do not take them into consideration when measuring the mean extension of subjects' future time perspective. We use them only for a qualitative description of the *temporal profile* (cf. p. 79).

The specific criteria for classifying an object in the "open present" category are explained in the *Manual of Time Perspective Analysis*. The same Manual also describes several additional symbols and combinations of symbols that are used in our coding system.

d. References to the past

Although our temporal code is primarily a coding system for the *future* time perspective, it does include some time categories for objects that refer to the past. It is, indeed, possible to refer explicitly or implicitly to past events when expressing motivational objects. Especially the negative inducer *"I am sorry (that)..."* (Nr n17) occasionally elicits references to past events, although responses such as *"I am sorry that I'll not be able to see my children at Christmas"* are more common (cf. *Manual*).

e. A temporal objects

Some objects cannot be localized in time. They are mostly pure fantasies (*I want to marry the moon*). This type of escape in unreal fantasies may be worth studying for itself, but we group them as unclassifiable (code "?").

V. RELIABILITY OF THE TEMPORAL CODE AND STABILITY OF RESULTS

To estimate the reliability of our temporal code, two trained judges code the data independently from each other, in the same way as for the content code explained above. Agreement between coders usually varies between 80 and 85 percent: 81 percent (Van Calster, 1971), 85 percent, (Verstraeten, 1974), 83 percent (Cossey, 1975), 85 percent (Nuttin & Grommen, 1975). Bouffard (1980) reports an agreement of 92 percent and 89 percent for two samples of MIM data. After the application of Cohen's (1960) formula of the K-coefficient, taking into account the percentage of agreement that may be caused by chance, Bouffard obtained a K value of .83, which approaches the maximal value of .96. He also recoded a sample of MIM data after a three-month interval and found a correspondence of 90 percent between the two codings.

MIM responses for which the two judges do not agree are discussed with a senior researcher. After such discussions, the agreement usually increases to 100 percent. The precise rules for coding time perspective as given in the *Manual* are partly the result of such discussion.

Verstraeten (1974) calculated the relative frequency of future time categories in MIM sentence completions of two aselect groups of 69 high school students (see Table 2). The correlation between the rank order of the time categories in both subgroups is .99.

Conclusion

The reader will have noticed that our temporal coding system is more detailed and differentiated than most time perspective classification systems. It gives a detailed picture of the temporal distribution of the subjects' motivational objects and their density in the various life periods. The time periods used have the advantage of being psychologically 'real' in the sense that they are also spontaneously used by people in describing the temporal localization of their life experiences. Besides its relatively high degree of differentiation, the main advantage of our technique is its 'objective' basis, which makes results of different studies highly comparable. The

TABLE 2

Relative number of motivational objects localized in the different temporal categories for two aselect groups of adolescents (E_2 life period; high school)(*)

Groups	D,W,M	Y,E_2	E,E_3	EA	A_1	A,A_2	AO	L	l	x
1	2.45	20.27	22.59	6.87	11.22	13.61	2.59	0.88	17.48	2.04
2	2.25	17.73	22.86	9.38	10.92	12.85	2.70	1.41	17.85	2.06

(*) For the meaning of compound symbols, see the Manual of Time Perspective Analysis.

psychological justification of this approach has been given. It must be recognized, however, that it does not belong to the category of 'quick methods'; it is time-consuming and some training is required for applying it correctly. It must be emphasized also that temporal localization of motivational objects is necessarily rather approximative because, most of the time, future behavioral events are only approximatively situated in the subject's mind. Measurement techniques cannot pretend to be more precise than the basic data allow for. Within these limits, the 'objective' classification on the basis of the temporal signs affecting behavioral events seems to be a safe technique for measuring the subject's future time perspective. As to the reliability of the technique, trained judges reach a degree of coding reliability that is sufficient for research purposes.

NOTE

(1) The term "object" is always used in its broadest sense. It can refer to an event, an object, a person, an animal, a situation, a relation, to anything that can be object of knowledge and motivation.

CHAPTER IV

MEASURING THE EXTENSION OF
THE FUTURE TIME PERSPECTIVE

The temporal coding of MIM responses as described in the previous chapter localizes a subject's motivational objects in one of the periods of his future. These temporally localized objects will now be used to measure the extension of his future time perspective (F.T.P.) and its density.

It is obvious that, for a given subject, the number of his future periods depends on his age (see Fig. 2). Our problem, now, is to measure the distance between the subject's present time and the time periods in which his motivational objects are localized. These distances are the basic for measuring the extension of his F.T.P. as a whole. Several methods will be proposed successively: the F.T.P. profile, the F.T.P. index, the median rank of F.T.P., and the mean F.T.P.-extension score. For the meaning of some symbols used in this chapter see the *Manual* of F.T.P. analysis.

I. FUTURE TIME PERSPECTIVE PROFILE (F.T.P. PROFILE)

The relative number of motivational objects localized in each of the temporal periods is calculated for homogeneous groups of subjects. The periods D, W, M, Y, followed by the periods E, A, O, and X, form a continuum, while the categories l, x, and L are special temporal units. Figure 3 shows two F.T.P. profiles based on the data in Table 2, p. 78.

There is, of course, no objection against adding the past category to this temporal profile. But the relative frequency of this category is generally very low.

The F.T.P. profile is used to *describe* the relative density of motivational objects in the more or less near future-time periods. For example, Table 2 and Figure 3 show that the present life period of the subjects has the highest relative frequency. Goal objects coded

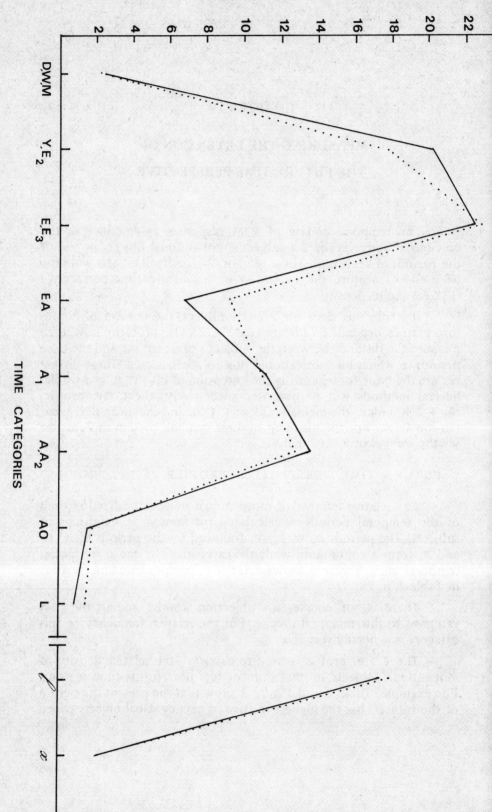

Figure 3. TWO SUPERIMPOSED F.T.P. PROFILES (BASED ON TABLE 2). The objects coded 1 or x form two special temporal categories.

E_2E_3 and even EA are situated in the current life period of the subjects (E_2 or E in general), although they extend into the next period (E_3 or A). A_1 objects are the first objects situated beyond the subjects' current life period. It also appears that the near future (D, W, M) and the category L have a very low frequency. The "open-present" (l), containing mostly goal objects related to personality characteristics and modalities, has a very high frequency. The F.T.P.-profile data, presented either as a table or as a curve, give a quick and descriptive picture of differences in the density of F.T.P. periods as a function of some experimental or differential variables.

II. FUTURE TIME PERSPECTIVE INDEX

The F.T.P. index expresses the proportion of motivational objects in the near future (D, W, M, Y, Y-) to the number of objects in the distant future (life periods). A higher relative frequency of more distant goals, in comparison with the number of immediate goals, expresses a longer future time perspective.

The time categories "open-present" (l) and "historical future" (x) are not used to calculate the F.T.P. index. Nevertheless, it may be important to compare the frequency of objects in the "open-present" with the frequency of references to the near and to the distant future. It was found, for example, in a study of the relationship between age and future time perspective, that the relative frequency of the open-present increased with age (until around 65) for a high socio-cultural group, while for lower social class people the relative frequency of the near future increased with age (see Nuttin & Grommen, 1975). Such a difference suggests an interpretation in terms of motivational contents that can be verified in a motivational content analysis of the MIM responses (*Ibid*.; Lens & Gailly, 1980).

III. THE MEAN FUTURE TIME PERSPECTIVE

In collaboration with some specialists in behavioral statistics (1), two techniques to calculate the mean temporal distance of motivational objects have been developed: the *median rank* of the F.T.P. and the *mean temporal distance* in number of years. Although it is fully recognized that expressing temporal distances of goals in mathematical terms is a delicate enterprise, the technique proposed

82   Extension of future time perspective

may have about the same symbolic meaning as many mathematical approaches to psychological phenomena.

A. The median rank of F.T.P.

The procedure is very simple: each of the successive time periods of our temporal scale receives a rank number in accordance with its distance from the present moment. The period D (day) receives rank 1, followed by 2 for W, 3 for M, 4 for Y, 5 for Y-. The series continues for the life periods. For each life period, three rank numbers are used in order to make a distinction between the beginning, the middle, and the end of that period. The differences in length between the time periods are not taken into account.

The time scale always starts with the current life period of the subject. The first part of that period is subdivided in short intervals in terms of calendar units. The immediate next interval following the last calendar unit (Y-) is the subject's current life period, and it always receives the rank order 6. For the end of this period, or when it is considered in its total duration, rank order 7 is scored.

The number of future time periods depends, of course, on the age of the subjects. A group of subjects in E_2 has four life periods (E_3, A_1, A_2, O); a subject in A_2 only one. For him, A_2 has the rank order 6, the end of that period (coded $A_2.$) or the whole A_2 period (code $\overline{A_2}$) ranks 7. The beginning of O (code $.O$) has the rank order 8, the middle old age, 9, and the end or the whole period ($O.$ or $\underline{O}$) 10, etc. Table 3 gives the rank number of the future time periods for different age groups.

The median of all the rank numbers of all motivational objects of a group of subjects can be used as an index of the average extension of the future time perspective of that group.

B. The mean extension of F.T.P.

This measure of F.T.P. extension is based on the length of the different time periods. The length of these periods is known more or less precisely. When classifying a motivational object in a certain time period, we usually do not know its exact place within that period. Sometimes it is possible to situate a goal in the beginning, the middle, or the end of a time period. Other objects extend beyond the

TABLE 3

Ordinal numbers indicating the rank order of each time category as a function of the subject's age (Ss)

Time categories of objects		Subjects of different age				
		Ss in E_2	Ss in E_3/A_0	Ss in A_1	Ss in A_2	Ss in O
Present life period calendar units		D = 01	D = 01	D = 01	D = 01	D = 01
		W = 02	W = 02	W = 02	W = 02	W = 02
		M = 03	M = 03	M = 03	M = 03	M = 03
		Y = 04	Y = 04	Y = 04	Y = 04	Y = 04
		Y- = 05	Y- = 05	Y- = 05	Y- = 05	Y = 05
beyond the calendar time		E_2 = 06	E_3 = 06	A_1 = 06	A_2 = 06	O = 06
		$E_2.$ = 07	$E_3.$ = 07	$A_1.$ = 07	$A_2.$ = 07	$O.$ = 07
Following periods:						
second:	beginning	$.E_3$ = 08	$.A_1$ = 08	$.A_2$ = 08	$.O$ = 08	$.X$ = 08
	middle	E_3 = 09	A_1 = 09	A_2 = 09	O = 09	X = 09
	end	$E_3.$ = 10	$A_1.$ = 10	$A_2.$ = 10	$O.$ = 10	
third:	beginning	$.A_1$ = 11	$.A_2$ = 11	$.O$ = 11	$.X$ = 11	
	middle	A_1 = 12	A_2 = 12	O = 12	X = 12	
	end	$A_1.$ = 13	$A_2.$ = 13	$O.$ = 13		
fourth:	beginning	$.A_2$ = 14	$.O$ = 14	$.X$ = 14		
	middle	A_2 = 15	O = 15	X = 15		
	end	$A_2.$ = 16	$O.$ = 16			
fifth:	beginning	$.O$ = 17	$.X$ = 17			
	middle	O = 18	X = 18			
	end	$O.$ = 19				
sixth:	beginning	$.X$ = 20				
	middle	X = 21				

An underlined symbol gets the same rank order as that symbol followed by a point. For subjects in O, the time period following their present life period is situated after their death (X). See the Manual of Time Perspective Analysis for the explanation of the different symbols in this Table.

limits of one period. We then use combined periods, such as $E3A1$ or AO.

Since we know the approximate duration of each period, we can calculate the approximate temporal distance between motivational objects situated in a period and the present moment. It is assumed that a motivational object of which we do not know the exact localization within a time period is situated in the middle of that period. Motivational objects at the beginning or at the end of a period have a corresponding temporal distance. To calculate the temporal distance of the different time periods we proceed as follows.

a) Calendar periods

The distance of objects that are localized within one of the short calendar periods is calculated in fractions of a year corresponding to their objective duration. Table 4 gives the details. These temporal intervals of the near future have the same temporal distance for all subjects (2).

TABLE 4

Temporal extension measure (in units of years) for each calendar interval

$D = 0.003$ (1/365 of a year)
$W = 0.02$ (1/52 of a year)
$M = 0.08$ (1/12 of a year)
$Y = 1$
$Y\text{-} = 2$

b) Social-clock periods

For objects situated beyond the calendar periods, the distance is calculated as a function of two variables: *the terminus a quo* is the subject's present age; the *terminus ad quem* is the age corresponding to the period in which the object is localized. The first variable is the real age of the subject at the moment of the MIM administration, or, as we usually define it, the mean age of the homogeneous group of subjects.

For the *terminus ad quem* we take the *middle point* of the age period in which the object is localized, unless the temporal code explicitly indicates the beginning or the end of that period.

First, the mean age of the group of subjects is calculated. To calculate the temporal distance of a motivational object, we apply the following rules.

Objects within the subject's current life period. These are objects that are situated further than two years from the present, but still within the current life period of the subject. We arbitrarily assume that they are situated in the middle of the remaining part of the current period.

Example: Objects that are coded A_1 for a subject of 30 year-old or for a group of subjects with a mean age of 30 are situated in the current life period of these subjects. In fact, subjects between 25 and 45 years of age are in their A_1 life period. As just stated, the objects coded A_1 must be localized in the *middle* of the *remaining part* of that period, i.e. in the middle between 30 (age of the subjects) and the end of their current life period (45), i.e. $(45 - 30) : 2 = 7.5$. This means that all A_1 objects are assumed to be localized at a temporal distance of 7.5 years for these subjects.

Objects coded A_1. or $\overline{A_1}$ are situated, for these same subjects, at the end of their current life period, i.e. at the age of 45. Their temporal extension value is $45 - 30 = 15$ years.

Objects in following life periods. All objects that are coded as situated in one of the life periods following the current life period of the subjects are assumed to be situated in the middle of that life period. Their distance is given by the difference between the age corresponding to that middle point and the real or mean age of the subjects.

Let us give the example of a motivational object coded A_2. It is situated in the middle between the age of 45 (the beginning of A_2) and 65 (the end of A_2), which is at the age of 55. It has a distance of $55 - 30 = 25$ years for a subject or a group of subjects who are 30 years old.

Objects situated at the end of a future life period (for example coded A_2. or $\underline{A_2}$ for 30 year-old subjects in A_1) have an F.T.P. extension value that is equal to the difference between the end of that period (65 for A_2) and the real or mean age of the subjects. Objects coded A_2. or $\underline{A_2}$ have a distance of 65 - 30 = 35 years for 30 year-old subjects. A motivational desire related to his death (coded $O.$, end of old age) has for a 60 year-old person an extension value of 75 (arbitrarily the end of life) minus 60, or 15 years (3).

For objects that are situated in the beginning of a future life period (for example: $.A_2$ for 30 year-old subjects), the distance is calculated on the basis of the age at which that period starts. However, because "the beginning" of a life period is not to be conceived as a moment, but rather as a more or less short time interval, we add 2 years to the age at which that life period is assumed to begin. The extension value or distance of goal objects coded $.A_2$ for 30 year-old subjects is the difference between (45 + 2) - 30 = 17 years. For the shorter social-clock periods, *viz.* E_2, E_3, and A_0 which have a length of only about six years, we add only one year to the age at which that period starts. An object coded $.E_3$, for a 14 year-old student (in E_2), is situated at the age of 18 + 1 = 19 and, hence, has a distance of 19 - 14 = 5 years.

For objects that *extend over two consecutive life periods* the same rules apply. However, two different cases can be distinguished: 1. the object extends over the actual and the following life period of the subject; 2. the object extends over two consecutive future life periods of the subject.

In the first case, the object is situated in the *middle* of the time interval between the present age of the subjects and the *end of the combined time period*. The difference between that middle point and the actual age of the subjects gives the extension value.

Examples:

A 20 year-old subject (in E_3) expresses a motivational object coded E_3A_1. That motivational object is situated in the middle (32.5) of the time interval between the age of 20 and 45 (the end of A_1), which is 12.5 years from the present [(45 - 20) : 2]

Another of his motivational objects is coded E_3A. It is situated in the middle between the age of 65 (the end of A) and 20 (the actual age of the subject) and has a temporal distance of:

42.5 - 20 = 22.5 years [(65 - 20) : 2].

A 30 year-old person (in A_1) expresses a goal coded AO. It is situated in the middle between the age of 75 and 30. Its temporal distance is also 22.5 years [(75 - 30) : 2] .

A motivational object coded A_2O for a 50 year-old person (in A_2) has a distance of 12.5 years [(75 - 50) : 2].

In the second case, the object extends over two (or three) periods following the current life period of the subject. The rules are the same. The object is arbitrarily assumed to be situated in the *middle* of the global period. Its distance is given by the difference between the age corresponding to that middle point and the actual age of the subjects.

Examples:

A 16 year-old subject (in E_2) expresses a motivational object coded E_3A_1. We situate the object in the middle of the time period between the age of 18 (the beginning of E_3) and the age of 45 (the end of A_1), which is at the age of 31.5. Its temporal distance equals 31.5 - 16 (the age of the subject), i.e. 15.5. An easy way to calculate the extension value of such objects is to calculate first half the length of the coded period (in our example: (45 - 18) : 2 = 13.5 years) and to add to this the number of years between the actual age of the subject and the beginning of the coded period (in our example 18 - 16 = 2 years). Hence, the temporal distance of E_3A_1 objects for a 16 year-old subject is [(45 - 18) : 2] + (18 - 16) = 15.5

A motivational object coded E_3A has, for 16 year-old subjects, a distance of: [(65 - 18) : 2] + (18 - 16) = 25.5 years.

Objects coded AO have, for a homogeneous group of subjects with a mean age of 20 years (E_3), an extension value of
[(75 - 25) : 2] + (25 - 20) = 30 years.

As a final example, objects coded A_2O have, for 30 year-old subjects, (in A_1) a distance of: [(75 - 45) : 2] + (45 - 30) = 30 years.

As to objects that extend over the total duration of a double period (for example code $\underline{AO}$), they are situated *at the end* of that total period (at the age of 75 for the $\underline{AO}$ code). Their temporal distance is simply given by the difference between the age marking the end of the double period and the actual age of the subject or group. For example, the distance of an object coded $\underline{AO}$ equals:

75 - 20 = 55 years for 20 year-old subjects.

We never use the code-symbol AO., which would refer to an object situated at the end of the combined adult-life and old-age period. If the object is situated at the end of a life period (here O), we code that period followed by a point (O.). When the object covers the whole duration of two successive life periods, we code the two periods and underline the symbols (for example $\underline{AO}$) (4).

For the symbol L we calculate half the distance between the age of the subjects and the end of life (arbitrarily set at 75 in our cultural group). For 30 year-old subjects, it has an extension value of $(75 - 30) : 2 = 22.5$. For 69 year-old subjects its extension value is only $(75 - 69) : 2 = 3$ (5). The code $\underline{L}$ (L underlined; see p. 114 for its meaning) has an extension value calculated by the difference between the end of life (75 years) and the age of the subjects. In the last example, it would be 75 - 69 = 6 years.

Objects coded X, referring to the time period following the subject's death, are arbitrarily situated 5 years later than the end of life, 75 + 5 = 80. It means that the X-period represents 10 years with its middle point 5 years after death. The beginning of that period (code $.X$) is situated 2 years after death (75 + 2). The motivational object of a subject (65 year-old) who is concerned about what will happen with his children shortly after his death (coded $.X$) has an extension of (75 + 2) - 65 = 12 years. A more general concern with the period after his death (coded X) is situated 5 years after the end of life and has for the same person a distance of 80 - 65 = 15 years.

Objects situated in the "open-present" (coded l) and objects situated in an impersonal time (coded x) cannot be localized in one of the life periods and are not used to calculate the mean extension of the future time perspective (cf. p. 75).

Two types of complex combinations of symbols can be distinguished when coding and measuring time perspective:

a) When the subject expresses a behavioral project with explicit mention of a goal object and of the means for reaching that goal, the two objects receive a temporal code. The two code symbols may be used for calculating the subject's F.T.P.

Example: A subject says: "I intend to work hard in high school in order to be admitted to a very good college". Our special code for this type of motivational objects is $E_2 \rightarrow .E_3$.

If, however, for some reasons it appears preferable to have only one code for each motivational item, the more distant element should be used for measuring the subject's F.T.P.

b) When one of the objects refers to the past and the other to the future (see *Manual* p. 103). the first is used to study the past time perspective and the second to measure the extension of the subject's F.T.P.

Note:

The mean future extension score can easily be used to compare the F.T.P. extension of groups of subjects of about the same age. But because it depends on the chronological age of the subjects, problems arise when the F.T.P. extension of different age groups are compared, especially when elderly people are involved. In fact, the objective length of the future is much shorter for an elderly subject than it is for his much younger grandchildren. It is, therefore, evident that older age groups will normally have a much shorter mean future-extension score than younger groups.

Because of this difficulty, Lens and Gailly (1980) proposed calculating for each subject of different age groups the proportion of his mean extension score to the number of years he is statistically expected to live. A high proportion means that the mean temporal distance of the motivational objects covers a relatively high proportion of the "objectively" to be expected life time of the subject and, hence, a relatively long future time perspective.

NOTES

(1) The senior author wishes to thank his colleagues at the Center for Advanced Study in the Behavioral Sciences in Stanford (1972-1973), especially Prof. Alan Stuart, for their helpful suggestions in this respect.

(2) A temporal interval in the very near future is subjectively experienced as much longer than a temporal interval with the

same objective duration but situated in the distant future or past (Cohen, 1967, p. 31). Our measure does not take into account such subjective perceptions.

(3) When subjects are afraid "to die before they are old", that object obtains a shorter extension value (for example, at the end of the adult life). When the individual says that he is afraid that he will die before a certain age, we use that age to calculate the temporal distance of this negative motivational object.

(4) We assume that a motivational object that extends over the periods A_2 and O (coded A_2O) has a shorter extension value than a motivational object that is situated in the O period only. An object coded O is situated at the age of 70 (halfway between 75 and 65), while an object coded A_2O is situated at the age of 60, halfway between 75 and 45, or:
$[(75 - 45) : 2] + 45$.
The same holds for A_1A_2 (or A) in comparison with A_2 only. An object in A_2 is situated at the age of 55; an object for A_1A_2 (or rather A) at the age of 45 $[(65 - 25) : 2 + 25]$. An object coded O is situated in the middle of the 65 - 75 period at the age of 70, while an object coded A_2O is situated in the middle of the 45 - 75 period at the age of 60.

(5) Although the symbol L refers to the total duration of the remaining life, we should recognize that this reference is rather vague. For example, someone who says *I hope to be successful in my life* refers to his whole life, but that time perspective is not necessarily explicitly present. That is why we calculate only *half* the length of the remaining part of life as an index of temporal distance, rather than the whole period. As such, objects coded L may have a shorter extension value than objects coded A_2 or O (see the examples given above for 30 years old subjects: $A_2 = 25$ years; $O = 40$ years; $L = 22.5$ years).

CHAPTER V

ATTITUDES TOWARDS
THE PERSONAL PAST, PRESENT, AND FUTURE

The content of past and future time perspectives are events, situations and projects that off and on occupy the individual's mind. These contents with their temporal sign (past or future) have an affectively positive or negative importance for the subject. It is because of their affective or motivational importance that the subject remembers or anticipates them from time to time (cf. p. 21). Neutral objects or objects without any importance are not integrated in an individual's time perspectives. The positive or negative characteristic (pleasant-unpleasant) of past, present, and future objects constitutes the affective attitude of an individual towards his personal past, present, and future.

These attitudes are important, not only because of their affective contents, but most of all because of their impact on the present motivation of the individual. For example, a positive or optimistic global attitude towards the future may increase the subjective value and probability of an anticipated motivational object and, hence, intensify the subject's motivation for it.

Attitudes towards the past, the present, and the future can be characterized by other components than their affective aspect. For example, the subject may have the impression that he is able to control his future or that it escapes him completely; he may perceive the future as very near or as still very far away, as exciting or as boring, etc. These components also affect the individual's motivation. We refer to Nuttin et al. (1979) for a brief theoretical and methodological discussion of time attitudes and attitudes towards time in general.

I. THE TIME ATTITUDE SCALE (T.A.S.)

The Time Attitude Scale (Nuttin, 1972) was constructed to

measure individual attitudes towards the personal past, present, and future.

The scale is based on Osgood's semantic-differential rating technique. We use this technique, however, not to measure the semantic content of the concepts past, present, and future, but to investigate the subject's attitudes towards these temporal dimensions and their content.

Based on preliminary studies and on their face validity, nineteen bipolar pairs of adjectives, such as *'pleasant-unpleasant'*, were selected. Some adjectives express an affective attitude, others refer to other aspects of a subject's motivational attitude towards the past, present or future (see Table 5).

Each pair of adjectives corresponds to a 7-point scale ranging from very positive to very negative. For example, for the pair *'pleasant-unpleasant'* the seven scale values are: *very pleasant, pleasant, rather pleasant, neither pleasant nor unpleasant, rather unpleasant, unpleasant, very unpleasant.* Subjects are asked to indicate on each scale how they spontaneously experience their personal past, present, and future. It is obvious that the present refers to the subject's present life period, and not only to the present moment in the strict sense. It is the period or situation experienced by the subject as his present condition. The 'present' includes the immediate future, as well as the recent past. It may be necessary, with regard to the purpose of a study or the situation of the subjects, to define the bounderies of the 'present life period'.

For measuring the attitude towards the past, only the first 15 pairs are used. For the present and the future, all 19 pairs are administered (see Table 5). The plus (+) sign following one of the two adjectives of a pair designates the positive pole. The most positive scale value is scored 7, the negative extreme is scored 1.

A few preliminary applications were made to study the characteristics of this scale. Lens (1972) administered the Time Attitude Scale (T.A.S.) to 260 male and female students in psychology and in educational sciences. He calculated the intercorrelation between the pairs of adjectives and discussed the results of factor analyses for each time dimension. The adjectives chosen on an a priori basis to measure the *affective* attitude towards the past, present, and future highly intercorrelated and formed a 'factor'. The total score for these pairs can be interpreted as a measure of the affective attitude.

These adjectives are the pairs 1 to 9 in Table 5. In some studies we included pair 11 (difficult-easy) and also the pairs 18 and 19 for the present and future. Lens found with 57 subjects test-retest reliabilities (after 6 months) of .44, .56, and .52 for the past, present, and future, respectively. De Volder (1978) obtained reliabilities of .74, .62, and .57 after an interval of three to four weeks with 198 first-year university students.

Van Calster's (1979) item analysis (Gulliksen-method) showed internal consistencies of .93, .94, and .92 for the attitude towards the past, present, and future, respectively (129 university students). He found for the same group of subjects a .70 correlation between the T.A.S. for the future and Goldrich's (1967) scale for optimism. With another group of 39 first-year university students, he found correlations of .47, .63, and .25 between the T.A.S. and Cantril's (1965) Self Anchoring Scale for the past, present, and future, respectively.

Thematic differences in temporal attitudes

The past, present, and future perspectives include motivational events and objects that occupy the individual's mind latently or actively. We may, therefore, expect that the attitudes towards the past, present, and future will be different when the subjects are invited to consider different categories of events or objects. For example, the attitude towards the past as far as family life is concerned may be different from the subject's attitude towards the past with regard to his professional career. The same is true for the present and the future. Instructions to express the attitude towards 'my future' in general are expected to arouse a global attitude in which the dominant elements cannot be identified.

To study the relationship between the global attitude towards each of the three time dimensions and the more restricted thematical attitudes, the T.A.S. was administered to 129 first-year university students. They were instructed to express their attitudes towards their personal past, present, and future: 1° in general (global attitude); 2° as far as their studies and/or professional life are concerned; 3° as far as their social contacts are concerned; and 4° as far as their personality development is concerned. The order of presentation of the global and the three thematic scales was randomized (cf. Van Calster, 1979).

Table 6 gives the mean scores for the global and the three the-

TABLE 5

The 19 T.A.S. pairs of adjectives to measure the attitude towards the personal Past, Present and Future (1)

	1 2 3 4 5 6 7	
1. Pleasant (+)		Unpleasant
2. Full (+)		Empty
3. Threatening		Attractive (+)
4. Beautiful (+)		Horrible
5. Cold		Warm (+)
6. Accomplished (success) (+)		Disappointment (failure)
7. Boring		Exciting (+)
8. Light (+)		Dark
9. Hopeful (+)		Hopeless
10. Fast (+)		Slow
11. Difficult		Easy (+)
12. Far away (+)		Near
13. Important (+)		Unimportant
14. Short		Long (+)
15. Externally determined		Something of myself (+)
16. Passively waiting for		An active task (+)
17. Staying unchanged		Always changing (+)
18. Open (+)		Closed
19. Familiar (+)		Unfamiliar

(1) The items 16 to 19 are used only for the present and the future. The items 1 to 9 are intended to measure the affective attitude. The numbers 1 to 7 at the top of the scale are used as reference marks for the subjects; they indicate the different degrees in which each adjective (pleasant, etc.) applies. When calculating the attitude score, an extreme negative answer is scored 1, and an extreme positive answer (see + sign) is scored 7. The + signs are, of course, not printed on the answer-sheet for the subjects.

TABLE 6

Mean affective attitude scores towards the past, the present and the future from four different points of view (T.A.S., items 1-9). Max. = 9 x 7 = 63

Time dimension	Thematic Points of view			
	Activities	Personality Development	Social Contact	Global
Past	44.47	44.38	46.67	45.17
Present	42.99	44.18	47.13	44.66
Future	45.58	46.39	49.05	46.30

matic attitudes towards the past, present, and future. The 9 pairs of affective adjectives were used.

Analysis of variance (with repeated measures for each of the two within-subject factors) showed a significant effect of the thematic factor [$F(3;381) = 23.82$; $p < .001$] and a significant effect of the factor 'time dimension' $F(2;254) = 7.54$; $p < .001$). The factor 'sex' (between subjects) did not have a significant effect, nor was there a significant interaction effect.

A posteriori Tukey Tests (Kirk, 1968, p.268) show that the affective attitude, as far as social contacts are concerned, is significantly ($p < .01$) more positive than each of the three other measures of affective attitudes. The data are combined over the past, present, and future time dimension because there is no interaction effect. A posteriori tests do not show significant differences between affective attitudes towards the past, present and future data combined over the four thematic measure because of no significant interaction.

The correlation between the three thematic attitudes and between the global attitude and each of the thematic attitudes are relatively high (see Table 7) and significant ($p < .01$). For each time dimension the highest correlation is found between the global attitude and the attitude towards personality development (.79; .86; .82).

Although there are differences between different thematic attitudes, it is also obvious that they are highly correlated with each

TABLE 7

Pearson correlation between different thematic attitudes and the global attitude towards the past, the present, and the future (N = 129)

	PAST			PRESENT			FUTURE		
	G*	A	C	G	A	C	G	A	C
Activities	.60			.56			.62		
Social contacts	.62	.31		.68	.37		.70	.60	
Personality development	.79	.58	.63	.86	.52	.73	.82	.72	.76

* G = Global; A = Activity; C = Contact.

other and with a global attitude. A common factor seems to be present in the different affective attitudes towards each time dimension.

Geirnaert (1976) compared the time attitudes of 50 neurotic military men and 50 physically ill or wounded soldiers, all hospitalized in the same military clinic. Leyssen (1974) applied the same T.A.S. to a rather heterogeneous group of 77 psychiatric patients with a mean hospitalization of 10 years and a mean age of 47 years.

Table 8 gives the results for these three groups and an aselect sample of university students (third and fourth year). The age of the latter group was comparable with the age of the soldiers.

The T.A.S. scores are grouped in three categories: *negative* (mean score below 3.50), *neutral* (mean score between 3.50 and 4.49), and *positive* (mean score of 4.50 or higher). Table 8 gives the relative frequency of subjects scoring in these three categories of affective attitude towards the past, the present, and the future.

The relative frequency of subjects with a negative affective attitude towards the past, the present, and the future is much higher in the two groups of mentally disturbed people than in the two other

TABLE 8

Relative frequency of positive, neutral and negative attitudes towards the past, present and future in different groups of subjects

	Psychiatric patients	Neurotic soldiers (hospitalized)	Physically sick (hospitalized) soldiers	University students
	N = 77	N = 50	N = 50	N = 260
PAST				
positive	61.0	44.0	82.0	73.8
neutral	15.6	36.0	10.0	21.5
negative	23.4	20.0	8.0	4.7
PRESENT				
positive	46.7	26.0	52.0	79.2
neutral	20.8	14.0	26.0	16.2
negative	32.5	60.0	22.0	4.6
FUTURE				
positive	63.6	66.0	98.0	91.2
neutral	16.9	18.0	2.0	7.3
negative	19.5	16.0	0.0	1.5

groups. The differences between the two groups of soldiers are also significant.

II. THE REVISED TIME ATTITUDE SCALE: A MULTI DIMENSIONAL SCALE FOR THE ATTITUDE TOWARDS THE FUTURE

The original T.A.S. included mostly items to measure the *affective* attitude towards the past, present, and future and only a few to measure some other aspects of attitudes towards the three time dimensions (Lens, 1972). The number of pairs of adjectives was too small to allow for an interpretation of these other aspects. In

TABLE 9

The revised T.A.S.: multi-factorial scale to measure the attitude towards the future (*)

Factor 1: "Structuration"

	saturation
Precise-Unprecise	.73
Well ordered-Chaotic	.70
Certain-Uncertain	.65
Structured-Unstructured	.60
Internal consistency (KR.20)	.85

Factor 2: "Internal Control"

	saturation
Planned by myself-Planned by others	.76
From myself-Externally imposed	.67
Depending on my efforts and/or capacities-Depending on luck or circumstances	.60
Personal-Impersonal	.46
Internal consistency (KR. 20)	.84

Factor 3: "Degree of difficulty"

	saturation
Difficult-Easy	.66
No conflicts-Full of conflicts	.63
Simple-Complex	.63
Unproblematic-Problematic	.48
Internal consistency (KR.20)	.75

Factor 4: "Value"

	saturation
Exciting-Boring	.65
Precious-Worthless	.60
Full-Empty	.59
Useful-Useless	.57

Internal consistency (KR.20)	.77

Factor 5: "Temporal Distance" (**)

	saturation
Near-Far away	.78
Immediate-Delayed	.77
Distant-Close	.74
Fast-Slow	.64

Internal consistency (KR.20)	.82

Global Affective Evaluation

Attractive-Threatening
Beautiful-Horrible
Pleasant-Unpleasant
Light-Dark
Warm-Cold

Internal consistency (KR.20)	.85

(*) The scales are translated from the original Dutch. On the answer sheets for the subjects, the scales can be printed as shown in Table 5.

(**) The factor "temporal distance" is related to the concept 'time perspective'. Distance or extension is the fundamental characteristic of that concept (see Chapt. I). Here, however, this factor refers to a general perception by the individual who sees the future as something still far away and who has the impression that time goes slowly.

order to construct a multi-dimensional scale, a member of our Center (Van Calster, 1979) extended the original list of adjectives with pairs of adjectives selected on a theoretical basis and related to other aspects than the *affective* component. He started with 58 pairs of adjectives to measure the attitude towards the *future*. The subjects were 236 first-year university students. Factor analysis showed that 66 percent of the common variance could be accounted for by one principal factor. Because the affective adjectives had the highest loading for that factor, we interpret it as the subjects' global affective evaluation of the future. Orthogonal varimax rotation resulted in five additional independent components in the evaluation of the personal future. In a non-identical replication with 230 12th grade students, the pairs of adjectives with the highest loadings in the first analysis revealed the same factors. The revised T.A.S. is, therefore, a multifactorial scale with five components in addition to the global affective attitudes.

Each of the five scales has four pairs of adjectives. The global affective attitude is measured with five pairs. Table 9 lists the six scales and their pairs of adjectives.

A multifactorial measure of the attitude towards the future makes it possible to study the *specific* effects of, for example, psychological disturbances on the attitude towards the future, as suggested by Lipman (1957) and Melges and Bowlby (1969). The relative importance of the different attitudinal components can also be measured.

It is obvious, however, that the factors found by Van Calster (1979) are probably not exhaustive. The factors resulting from factor analysis are limited by the content of the items used.

CHAPTER VI

MANUAL OF TIME PERSPECTIVE ANALYSIS

by J.R. NUTTIN & W. LENS

I. CODING UNITS AND THEIR CONTEXT

II. DESCRIPTION OF THE CODE

    2.1. The symbols
    2.2. The meaning of the symbols: the temporal categories
        2.2.1. Calendar periods
        2.2.2. Social and biological life periods
        2.2.3. The total duration of life
        2.2.4. The 'open-present'
        2.2.5. A time period after death and the historical future
        2.2.6. References to the past

III. THE CODING TECHNIQUE

    3. 1. The subject expresses the temporal localization
    3. 2. The importance and nature of the motivational object
    3. 3. Social or biological links with a life period
    3. 4. The arrangement of calendar periods and life periods
    3. 5. The total life period
    3. 6. The 'open-present'
    3. 7. The historical future
    3. 8. Combined temporal periods
    3. 9. Additional specifications of symbols
    3.10. Repetitive acts or goal objects
    3.11. References to the past
    3.12. Negative goal objects
    3.13. Motivational objects for others
    3.14. Atemporal objects and uncodable responses
    3.15. Global time categories

IV. LIST OF EXAMPLES

CHAPTER VI

MANUAL OF TIME PERSPECTIVE ANALYSIS

by J.R.NUTTIN & W.LENS

In the preceding chapters, a theory of time perspective and a method for measuring its extension have been introduced. The present *Manual* is intended as a practical guide for localizing the subjects' "motivational objects" - collected by the MIM - in the time periods in which they are expected to happen or to be realized. These time periods, it will be recalled, are the temporal elements that constitute a subject's future time perspective. The principles of our temporal coding system and the temporal scale to be used have been described in Chapters I and III, respectively. The reader is to keep them in mind before using this Manual in which the concrete rules for localizing the objects are presented. It is on the basis of these temporally located objects that the extension measures of the subjects' Future Time Perspective (F.T.P.) are obtained, as described in Chapter IV. The Manual for the motivational content analysis of the MIM responses is given in the next chapter. This content analysis, however, is not necessary for measuring the subjects' F.T.P.

Unlike most other techniques for measuring time perspective, our temporal coding system allows for a strongly differentiated temporal localization of the objects the individual has in mind. This is an important advantage, but it also poses many practical difficulties and problems.

It will be recalled that our temporal scale is divided into two rather heterogeneous parts: one part with calendar units and one with units in terms of the social and biological clock (see Chapt. III). It may be necessary to make some adaptations in the second part of our scale for subjects living in other cultures. Moreover, some types of motivational objects cannot be localized within a specific time period: Their temporal location requires special temporal categories and code symbols, as will be explained later.

It should also be recalled that our temporal code is based on what we have called an 'objective' or 'average' time location of objects and events (cf. p. 68). The question to be answered for each motivational object when coding its time location can be formulated as follows: 'Assuming that the life of this person has a normal course as for most of the people of his group, when in his lifetime will this motivational object be obtained or realized, or when is this event most likely to happen?' For negative motivational objects the question to be answered is 'In which time period is this negative motivational object most likely to really threaten the subject?'. The temporal localizations resulting from answering these questions are expressed in temporal code symbols (see below). The coder can proceed as follows (practical rules are numbered from 1.1. to 3.15 for the sake of convenience in references).

I. CODING UNITS AND THEIR CONTEXT

Before coding the time location of each individual answer, the coder should read the whole booklet of a subject's sentence completions. By doing so, he becomes familiar with the context in which each sentence should be understood. That context may affect the meaning of a particular answer and, hence, its localization in time.

1.1. Each motivational object expressed in the MIM sentence completions is considered as a unit for analysis and coded as such. 'Coding' here means the attribution of a temporal code symbol to each motivational object, the symbol indicating the localization of that object in its estimated time period. This estimation is done according to the rules formulated in this Manual.

1.2. A sentence completion in which, exceptionally, no classifiable object is expressed receives the symbol "?" (see p. 131 for examples).

1.3. When two or more motivational objects are expressed in one sentence, coding should be done according to a uniform rule decided on in advance. If it was decided to code not more than one object for each inducer, the object with the longest future extension value should be coded. Indeed, the object is to measure the extension of the subject's future time perspective. If, however, the special purpose of a research project requires all objects to be taken into account in order to study, for instance, the density of each time period in a subject's future time perspective, each object should receive its own code symbol. Example: *I hope...to finish my studies, to find a decent job, and to marry.*

The same rule applies when the subject expresses two or more alternative and mutually exclusive goals. Example: *I want...to go to college or to find a job; I intend to...go to the football game or to see some friends.* In many such cases, however, the two alternatives may have the same temporal localization.

If the subject makes a distinction between the importance of two or more objects expressed in one sentence completion, it may be desirable to code the most important object when it was decided to code only one. Example: *I hope...to finish my studies, but most of all to find a decent job and to marry.*

When two or more objects are coded, the symbols for the different objects are separated by a plus (+) sign and have the same order as the objects in the sentence completion.

1.4. When a means-end structure is expressed, both elements (the means and the end) are coded and the two symbols are connected with an arrow (→) to express the means-end relationship. Example: *I hope...to be very successful in college in order to have a nice career* (= $E_3 \rightarrow A$). The two symbols are taken into consideration when calculating the frequency of the different temporal categories and the mean future extension score. Even when it was decided to code only one element in cases such as mentioned in 1.3, it may be desirable that both elements of a means-end structure be coded with the special arrow symbol in order to allow for studies of the internal structure of the subject's time perspective. If, however, only one object is taken into account, the goal object should be coded and not the means.

1.5. Several temporal nuances and modalities expressed in the subject's sentence completions are taken into account (e.g. *sometimes, much later, always*, etc. see p. 117). The same holds for temporal modalities implied in verbs such as *to become, to stay*, etc. Example: *to become a psychologist* refers to the subject's educational period at the university level (code = E_3), whereas *to be a psychologist* refers to the professional career (code = A).

1.6. The meaning implied in the inducer itself is never coded. The sentence *I am working towards...a Ph.D.* is not coded as a means-end structure because the first part of the sentence is the inducer. We use the inducer only to understand the meaning of the sentence completion itself.

Example 1: *I wish very much to be able to...swim.*
Example 2: *I wish...to go for a swim.*

In the first example, the subject wants to have a certain *capacity* that applies already in the present, but that also goes on indefinitely in the future. That category of objects belongs to what we have called the *open-present*, its code symbol being the small letter *l*. In the second example, he wants to perform an *act* that usually can be realized either within a day (code: D), or on one of the following days (code: W) according to the subject's circumstances.

1.7. As mentioned above, the correct understanding of a particular response, or even its general meaning, may require an interpretation in the context of a previous response or of the subject's booklet as a whole. Example: *I hope...to succeed.* For a subject who referred several times to his final examinations at the end of the academic year, it may be safely assumed that he is talking about success in his exams (code Y.). But the temporal coding should be different when previous responses make it plausible that he is talking about his professional career as a whole (code A) or of a competition to be held in the coming weeks (code M).

1.8. All available information on the group of subjects or on an individual subject (age, sex, marital state, profession, social class, etc.) can be important in the understanding of the motivational objects. 'Success' does not refer to examinations when expressed by a young adult starting his professional career (code A). For a group of young athletes who must go through a difficult selection at the end of the week, 'success' will refer most probably to that test at the end of the week (code W). The sentence *I want...to be with my family at home* has a different temporal localization when expressed by a college student who goes home every week than when expressed by a prisoner recently sentenced to a 20-year term. In general, information about the individual is more important than information about his group when interpreting the meaning of individual responses. All necessary information is asked for and given by the subject on the first page of his personal booklet.

II. DESCRIPTION OF THE CODE

2.1. *The symbols*

The different types of symbols used in our temporal code are enumerated first; their meaning is given in the following section in terms of temporal categories.

2.1.1. Capital letters (upper case) referring to the time periods
   a. in terms of calendar units: *T, D, W, M, Y*
   b. in terms of social and biological life periods: *E (E_2, E_3), A (A_0, A_1, A_2), O, X, L, P*

2.1.2. Small letters (lower case) referring to more vague temporal durations and localizations: *l, x, p*

2.1.3. Combined symbols
   a. two or more capitals: *EA; AO;*
   b. a letter followed by a capital between parentheses specifies the localization in the "open present" or in the past: *l(E); P(A_1);* etc. (cf. infra).

2.1.4. Additional signs can be added to the capital letters:
   a. the numbers 0, 1, 2, 3 added to the capitals *A* and *E*: E_0, $E_1, E_2, E_3, A_0, A_1, A_2, E_3A_1$
   b. a dot before, under or after a capital: $\dot{E}_3; \dot{A}_1; O.; .E_3; .A; .O;$ etc.
   c. a dash before, under or after a capital: $\underline{A}; \underline{L}; Y$-; -O
   d. arrows to express relationships between two capital letters: ←; →; ↔;
   e. a question mark is coded when the response cannot be understood or when it is a pure fantasy (symbol: *?*)

2.2. *The meaning of the symbols: the temporal categories*

A capital letter without a specification refers to a temporal period in its totality. A motivational object coded with a capital is assumed to be situated somewhere in that period. It may occupy a longer or shorter time interval or be limited to a transitive activity (a trip, for example) happening once or regularly in that period (cf. p. 118). Additional signs are added to the capital letter to indicate that the object or event must be situated in the beginning or at the end of the period, or that the total period is to be taken into account (see p. 117). A *dash* after a *capital* (mostly M- and Y-) refers to a prolongation of that period (two or three months; one or two years).

2.2.1. Calendar periods (see also p. 75)

T	the present moment, the duration of the MIM application
D	within one day
W	within one week
M	within one month
M-	within two or three months
Y	within one year
Y-	within one or two years

2.2.2. Social and biological life periods

The calendar-time periods are situated in the present life period of a subject. Motivational objects that are localized beyond the calendar time (i.e. beyond Y-) but still belong to the present life period of a subject are coded with the capital letter indicating that present life period (see below Nr 3.4.). Thus for a high school pupil, the symbol E_2 refers to his motivational objects situated within his high school period but beyond an approximately one-year interval.

E (educational period). The more specific symbols E_2 and E_3 should be used instead of E wherever possible. For example, the combined symbol E_2E_3 is preferred to E for coding the pupil's objects that span the whole high school and post-high school education period.

As will be shown later, the E-symbol is used between parentheses after the symbol l (open present) when an object in the open present is limited to the educational period, and also after the capital letter P (the Past) to indicate that an adult or older subject refers to his educational period.

E_0 This symbol refers to the first years of life (before first grade). We use it only for MIM sentence completions that refer to that past life period of a subject (code $P(E_0)$).

E_1 This symbol refers to the primary school period (6-12 years old). Because the MIM cannot be used for this age group, the E_1 code can only be used, in the same way as the E_0 code, for memories (code $P(E_1)$).

E_2 This symbol is used for objects situated between the age of 12 and 18 (junior high school and high school).

E_3 The period of higher education (18-25 years). The code A_0 is substituted for E_3 when there is no period of higher education in the subject's life.

A_0 This symbol corresponds to the intermediate period between 18 and 25 years of age for subjects who do not continue their education and who are not yet adults because of their family or professional situation. Some adaptations may be required by the socio-economic conditions of subjects in other cultures.

A Adult life is assumed to begin - at least in the Western world - when the subject becomes independent of his parents and has his own family life or professional and economic career. Marriage and/or earning his own living are the most usual criteria for it.

The symbol A is coded when the motivational object is situated in the A period and when a more precise temporal localization within that long period is not possible.

A_1 For objects situated in the first half of the adult life (25-45 years).

A_2 For objects situated in the second half of the adult life (45-65). Such objects or goals refer mostly to the peak of a career or to a more advanced stage in the professional or family life.

AO The motivational object is situated in the A-period but also in old age.

O The object is situated in the period of old age (starting with professional retirement or at the age of 65). As mentioned earlier, the age at which the different life periods start may be culturally determined. Some adaptation may be necessary for different cultural groups.

2.2.3. The total duration of life

In some answers the subjects refer more or less explicitly to the total remaining part of their life. We distinguish two types, coded L and L̲:

L   Capital L is coded when the reference to the total (remaining) life period is only vague and not very explicit (example:...*to succeed in life*).

L̲   Capital L with a dash under it is coded when such a reference is very explicit (example:...*to devote my whole life to the poor*).

2.2.4. The 'open-present': *l* and *l̲* (cf. supra, p. 75)

l   The motivational object is a modality of being, a quality or a capacity that the subject wants for the present moment, but also for a more or less undefined future. The goal must be situated in, or applicable to, the present but it refers implicitly also to the subject's future lifetime or some part of it. When the quality or capacity does not apply to the individual's present situation (for example:...*to be a good mother for my children*... expressed by a girl or woman who does not have children), the capital letter of the life period corresponding to the goal object is coded (here *A 1*).

    We do not use the code *l* when the subject talks about exercising - even habitually - an activity. Only the *capacity* to do something is coded *l* (*I would like so much...to be able to speak French fluently*). Moreover, when the individual wants to *acquire* (or to *learn*) some capacity or to *become* something (e.g. more educated) the time period needed for that preparation or development is coded (with capitals) and not *l*.

l̲   The dash under the symbol *l* is used when the subject stressed the prolonged duration of the quality wanted by using terms such as 'always', 'to stay', etc., in connection with the quality or modality of being (examples on p. 129).

l(E)   A capital letter between parentheses and following *l* means that the future implied in the open present is limited to the period between parentheses. The present is open to the future but not further than the life period coded between parentheses.

2.2.5. A time period after death and the historical future

X   The motivational goal is situated after the subject's death, but it is still related to himself, his family, or his personal interests (see also p. 128 and p. 129).

x   This symbol (small or lower case *x*) is used for humanitarian or rather general motivational objects that cannot be situated on an individual's life scale. They rather belong to the development or evolution of history (see p. 129).

2.2.6. References to the past (P)

P   The object or event expressed in a MIM sentence completion is situated somewhere in the individual's past. Example: *I am sorry that I was so lazy.*

P(...) If we know the past life period in which the object must be situated, we code the corresponding capital in parentheses. For example: $P(E_3)$; $P(A_1)$. Example: *I find it unbearable that...I was so lazy in high school* (code $P(E_2)$).

When the motivational object is situated in the future, but refers in an implicit or explicit way to a past event or situation, the two temporal dimensions are coded with an arrow from the past code to the future code. This type of coding is explained below (see from 3.11.1. to 3.11.3.).

III. THE CODING TECHNIQUE

As stated above, our temporal code allows for a much differentiated temporal localization of motivational objects. Moreover, we do not direct the subject's attention towards a certain future period by asking, for example, *What will you be doing in five years from now?* For these two reasons, localizing motivational objects on our temporal scale is a difficult enterprise. A certain lack of precision can sometimes not be avoided and is inherent in the temporal localization technique (cf. above, p. 78).

The following practical rules provide the coder with safe guiding lines. Each rule is illustrated with one or two examples. A long list of examples is to be found at the end of this chapter (Section IV).

3.1. *The subject expresses the temporal localization*

3.1.1. Sometimes a subject expresses in a sentence completion not only a motivational goal but also its temporal localization. In such cases we code the subject's temporal localization of the goal.

Example: *I hope...to buy a car this year* (code *Y*).

3.1.2. When the subject's temporal reference is unreal or impossible a double code is given: the 'mean' or average temporal localization of the object, followed by the code for the subject's localization. The two codes are separated by a double arrow (↔) to indicate the conflict between them.

Example: *I would like so much...that summer vacation would start tomorrow*: Coded *Y↔D* when vacation will begin only after several months. When the vacation will begin in a few weeks, we code (*M↔D*). For all further analyses and calculations the two codes are used.

3.1.3. When the subject formulates a motivational goal that is related to his present activity of answering the MIM (*I will be glad when... this test is finished*) the symbol *T* (test) is coded.

3.2. *The importance and nature of the motivational object*

The more or less important or trivial nature of a motivational object will often help to situate it in a more or less distant time interval. Daily activities and the satisfaction of certain physiological needs (eating, sleeping, etc.) are situated in the very near future.

Examples:*...to have a good night's sleep;...a good dinner;...to smoke a cigarette*, are motivational objects to be situated within one or two days (code *D*); *...to take a long walk;...to see a friend; ...to go for a swim*, are most probably to be situated within a time interval of a few days or a week (code *W*).

The importance or difficulty of some motivational objects is also related to their temporal localization. *To buy a new car*, or *to build a house* takes usually more time to be realized than *to buy a bicycle*. The social situation of the subject is to be taken into account in order to determine whether 'buying a car' is a question of a few months (*M-*) or a few years (*Y-*). Most groups of youngsters can only buy their own car when they have a job (code $.A$ or $.A_1$).

3.3. *Social or biological links with a life period*

Many goal objects are by their very nature associated with a certain social or biological life period. This is true, for example, for all goals that are related to formal education (in primary or secondary school, college or university). All motivational aspirations and fears about a professional career or retirement are closely linked to one (or two) life periods of our temporal scale.

Examples: Finding one's first full-time job is situated in the beginning of the adult life ($.A$).

Raising a family takes place during the first half of the adult life (A_1).

Professional life covers the whole duration of the adult life (A).

An important professional position or the peak of a professional career is usually not reached before the second half of adult life (A_2).

Family life spans adult life and old age (AO).

Retirement starts with old age ($.O$) and coincides with it (O).

The personal death marks the end of life ($O.$). A subject who is, for example, concerned about what will happen to his children after his death, has a motivational concern situated after his death ($.X$ or X).

3.4. *The arrangement of calendar periods and life periods*

For each subject, future time perspective (and past perspective) starts with periods defined in terms of calendar units: the time categories D, W, M, Y, Y-. The social or biological time categories (E_3, A_1, etc.) are only coded when the object cannot be situated within a one or two years period (Y or Y-).

Objects that are situated beyond an interval of one or two years (Y-) but still within the current life period of the subject (for example A_1), are coded with the symbol for that life period (here A_1). For a subject who is in the first half of his adult life, code A_1 means that the goal object cannot be realized within the first two coming years but certainly during his present life period (before the age of about 45). This is true for all codes that correspond to the current life period of the subject. For some subjects, for example students in the final year of their school education, an interval of two years is longer than the remaining part of their current life period and extends into the beginning of the next period. In such cases, we also code in terms of calendar periods (Y- rather than for example A_1 (see however note, p. 127). As explained above, the temporal distance implied in a code depends on the age of the subject. The symbol A_1 has a longer future extension for a subject in E_2 (high school) than for a college student (E_3).

3.5. *The total life period*

3.5.1. When the subject explicitly mentions 'my life' or 'my existence', the motivational object is coded L. We do the same when the term 'ever' or 'always' is used in a context that implies, at least implicitly, the whole remaining life period (Ex.: *I hope...that I can always stay in this town*). However, when the concept 'life' is used without the implication of duration (...*to live a modest life*) the object is situated in an 'open-present' (code *l*). The subject wants to continue a habit. It is not always easy to make the distinction. In such cases, we should remember that objects coded *l* are first of all related to the present situation that one wants to continue in one's present state. To calculate the extension value of objects coded L, see p. 88.

3.5.2. $\underline{L}$ (with a *dash* under the capital letter). The symbol $\underline{L}$ is coded when the subjects refers more explicitly to the total duration of his whole remaining life by using expressions such as 'every day of my life', 'my whole life long'.

3.6. *The 'open-present'*

The meaning of the 'open-present' category is explained in Chapter 3 (cf. p. 75). A rather large number of motivational goals expressed in MIM sentence completions have to do with a state of being, a quality of the person, a personality trait, an aptitude, a physical quality, a general disposition. They are to be situated in the present but at the same time in a vaguely envisaged future. The subject's intentional look is oriented toward the present, but at the

Coding technique 115

same time extends into the future, although not very explicitly.

In coding objects as belonging to the category 'open-present', the following rules will be helpful:

3.6.1. Wishes for a personality trait or a certain modality of being (e.g. to be beautiful, intelligent, punctual), a capacity or aptitude (to speak several languages, to be able to write computer programs, to know how to swim, etc.) belong to the 'open-present'. We have already mentioned the difference between the wish 'to be able to swim', as a capacity one wants for now and later on (coded l), and 'to go swimming' as an act that can be performed today or in a few days.

3.6.2. The continuation of a situation (see paragraph 3.1.3. and 3.6.3.).

3.6.3. The symbol l is followed by a capital letter between parentheses when the quality or modality of being is limited to the subject's present life period or to that period and one or more of the following periods. For example: *I want to be a hard-working student* is coded $l(E)$ for a student-subject. Its content applies now and during the remaining part of the E-period, but not beyond it. '*I hope to be a good mother for my children at home*' refers to the first half of the adult life (A_1). When a young mother (in A_1) expresses this object it is coded $l(A_1)$. It is coded A_1 when expressed by a subject who is not yet a mother. This person thinks about her future and not about a quality she wants to have from now on. The capital letter between parentheses and following l refers always to the present life period of the subject (and sometimes its prolongation, for example, $l(E_3 A_1)$ for a subject in E_3).

The sentence completion '*I hope...that my parents will continue to stay with us*' is coded $l(A_1)$ for a subject in A_1. In this case, the goal object is related to other persons but is situated on the temporal life scale of the subject and not on the life scale of those other persons.

3.6.4. The category 'open-present' is not coded when the motivational goal does not apply to the current situation of the subject. In such cases, the subject is thinking about the life period in which the object does apply. The capital letter for that period is coded. An unmarried high school student who says '*I intend to be loyal to my wife*' cannot yet be loyal to his wife because he is unmarried. We do not code *l*, but the temporal period during which he will be married (*AO*). In general, when the motivational object is formulated in terms of change or development, we code the time interval required to obtain the quality or characteristic. Example: *I hope...that I'll be able very soon to speak fluently French* (coded *Y-*) (cf. par. 3.6.5.).

3.6.5. The category 'open-present' is not used either when the motivational object is expressed in terms of an activity. For activities that are already going on, the temporal periods in which they happen are coded (Example: *to study with diligence* is coded E_2 or E_3 (or *E*) according to the subject's age). It is evident that verbal expressions that explicitly refer to the future, such as 'very soon', 'one day', 'later', etc., exclude an 'open-present' coding. Example: *I hope to be rich one day* is coded A_2 and not *l*.

3.6.6. Whenever a sentence completion mentions or implies a process of 'becoming' (I hope...that I *will be* a good mother for my children), or when the inducer explicitly refers to the future (*I will be very happy when...* Nr 25), the motivational object is situated in the time interval during which the development or change happens or in which the object will be realized by the subject. We use the capital letter for that period.

3.6.7. The code $\underline{l}$ (dash *under small letter l*) is used when the duration is stressed. For example: *to be happy for ever; never to lose the confidence of my wife*, etc.

3.7. *The historical future* (symbol *x*)

The code *x* is used for motivational goals that cannot be situated on an individual life time scale, but that are related to a historical future. Examples: *I want...freedom for all people; I would like...more social justice in the world*. The motivational object refers not to the subject's personal life but to mankind, society, communities, peoples, etc. (see p. 129 for examples).

3.8. *Combined temporal periods*

3.8.1. When a motivational object is not limited to one temporal interval but extends over two or more, the capital letters of the different time periods are combined. When a college student who lives with her mother says *'I am determined to...take care of my paralyzed mother'*, we code E_3A_1. We assume that this student intends to do this as long as her mother lives, that is most probably during the subject's E_3 and A_1 life periods. The code AO, for example, is used for objects that cover the adult life and old age: *I try to...help my wife at home*. Another example of combined periods is E_2E_3. We usually do not code A_1A_2 but simply A to indicate that the total adult period is implied.

3.8.2. When two (or more) motivational objects are expressed in one sentence completion, different coding rules are possible (cf. supra paragraph 1.3.). When it is decided to analyze all the responses, the temporal codes of the different responses are connected with a 'plus' (+) sign, and all the codes are used for further analysis. When it is decided to code not more than one object in each sentence completion, the aim of the study should determine which one is coded. When studying the extension of the future time perspective, the most distant goal object should be coded. When a college student says *I am determined to...be successful in my studies and in my professional career* we code $E_3 + A$ or only A.

3.9. *Additional specifications of symbols*

3.9.1. A dot *before, after*, or *under* a capital letter

To accentuate the fact that the motivational goal or event will happen only once or occupies only a short time within a certain temporal period, we code a *dot* under the capital letter for that period (for example $A_{\dot 1}$). But unless the goal of a study requires making a distinction between an event of short duration and an object or event that applies to a longer part of the period, we do not use the code 'dot under capital'. Indeed, the future extension value of both codes is the same. In both cases (A_1 and $A_{\dot 1}$), the object is assumed to be situated in the middle of the A_1 period (or in the middle of the remaining part of that period), as explained in Chapter 4.

A dot *following* a capital letter ($A.$) is used when the goal is normally situated at the end of the period referred to by the capital letter. A dot *before* a capital letter refers to the beginning of that period. The code $O.$ refers to the end of life. The student who hopes to find a job after college has a motivational goal for .A (or .A1). His goal *to obtain my doctoral degree* is coded E3..

3.9.2. *A hyphen after* a capital letter is coded when the object or event may extend a bit beyond the period referred to by the capital letter (Y- = within one or two years). *A hyphen before* a capital letter means that the object is situated in the period referred to by the capital letter but may appear already a short time before that period (e.g.; -O). We rarely use that symbol.

3.9.3. *A dash under* a capital letter is coded to indicate that the object applies to the total period, or that it may be repeated several times during that period (Example: $\underline{A_1}$) (cf. infra:par. 3.10).

3.9.4. *A dash under small letter l* is coded when the subject stresses the duration of the object situated in an 'open-present'. When the sentence completion explicitly refers to the total life period, capital L is coded. Examples of $\underline{l}$: I want to be happy *for ever*; I would like so much *to stay* healthy.

3.9.5. An *arrow* (→)between two capital letters means that the two goal objects form a means-end structure. For example *I intend to...go to another university in order to find a job* is coded E3→.A. An arrow headed to the left (←) is coded when the expressed goal refers to the past (see under 3.11.).

3.9.6. A *double arrow* (↔) indicates a conflict between the temporal localization given by the subject in the sentence completion and the objective time localization (cf. supra par. 3.1.2.).

3.10. *Repetitive acts or goal objects*

Subjects often express motivational objects that they want to perform repetitiously or at fixed intervals. They want to go to the theatre *regularly*, to go hiking *every weekend*, to visit their parents *every year at Christmas*, to play at cards *very often*, etc.

It is not easy to grasp how far into the future the intentional look of the subject reaches when formulating such goal objects, so we need some practical rules that are helpful for a uniform coding of their time perspective.

3.10.1. When the repetitive act or object is implicitly or explicitly situated within a certain time period, we code the capital letter for that period with a dash under it. For example *each day of my life* is coded $\underline{L}$; when a student says that he wants 'to go home *every weekend*' we code $\underline{E}_2$ for a high school student and $\underline{E}_3$ for a student in college or graduate school.

3.10.2. We assume that objects of some importance that are regularly repeated, but only after long time intervals, imply a longer future time perspective than less important objects that are repeated more frequently. It seems reasonable to assume that the businessman's intention to change his car *every three years* implies a longer future time perspective than his intention to go to the movies *every Saturday*. With his first goal he has probably his whole professional career in mind, although perhaps only vaguely (code $\underline{A}$). Less important decisions, such as his second goal, or such as 'to take a short walk every morning', 'to go to the library very often' or 'to play cards once a week' are assumed to have a much shorter future time perspective no longer than one or two years (code $\underline{Y}$-). A motivational project with an intermediate importance or frequency of repetition, such as the young adult's intention to visit his parents each year at Christmas, is coded $\underline{A}_1$.

3.11. *References to the past*

Although motivational goals are obviously localized in the future and, therefore, cannot be used to study the subjects' preferential orientation towards the past, it does happen that subjects refer to past events while formulating motivational objects and concerns. We distinguish the following three types of references to the past.

3.11.1. A motivational object or concern is sometimes to be localized in the past. This may happen especially for regrets and for sentence completions to negative inducers. Example: *I think it is sad that...I did not finish high school*.

Such motivational concerns are coded with capital P (Past). If possible, the Past period that is referred to is mentioned between parentheses following P. For the example given, the code is $P(E_2)$. The response *It displeases me that...I invited our neighbors last week* is coded $P(W)$.

3.11.2. Besides the motivational goal that is to be realized in the future, the subject *explicitly* refers at the same time to an event or situation in the past. Example: A college student (or a recently married man) says: *I intend to...give my children what I missed as a child*. The goal object refers to his future children and is situated in his A_1 life period; but he also refers to his own childhood experience (E_0E_1 life period). The two temporal codes are given: first the future-code A_1, and second the past-code $P(E_0E_1)$. The two codes are connected by an arrow from the past to the future. The complete code being $A_1 \leftarrow P(E_0E_1)$. It should be noticed that only explicit references to the past are coded with a capital letter P.

3.11.3. *Implicit* references to the past are found in expressions such as 'no more', 'not again', etc. Example: *I intend to... never smoke again*. The motivational goal is: *not to smoke anymore, neither in the present nor in the future*. Taking into consideration that the subject refers to an activity (smoking) and that he uses the expression 'never', which implicitly refers to the remaining part of his life, we code L. But 'never again' also means that the subject did smoke in the past. It is an *implicit* reference to the past, which is coded by the small letter p. The complete code for this example is hence: $L \leftarrow p$.

When the subject explicitly refers to his whole future lifetime, the symbol L is underlined ($\underline{L}$). Example: *I intend to...never smoke again in my whole life* (code $\underline{L} \leftarrow p$). Small letter p is not followed by any specification. Without the implicit reference to the past (never...again) the code would be $\underline{L}$.

Although one may consider it as a matter of unimportant nuances, differences in verbal expressions correspond to real differences in the content of cognitions. Also expressions such as 'to change my life' or 'to mend my ways' refer implicitly to the past. As said above, expression such as 'to become', 'to develop', etc. refer to the future, and we therefore code the future time interval during which the development will take place. Such differences in verbal expression are also taken into consideration when coding the motivational *content* of MIM sentence completions (cf. Manual of Motivational Content Analysis).

3.12. *Negative goal objects*

Goal objects expressed in sentence completion to negative

Coding technique   121

inducers are coded following the same rules as when coding positive motivational goals. Here also we answer the question 'when is the negative goal object actually threatening an average subject of this group of people?'

We distinguish the following categories of responses that are frequently expressed to negative inducers.

3.12.1. The subject regrets that he has done (or not done) something in the past.

Examples: *It displeases me that...I did not finish high school*: $P(E2)$
I am afraid that I married much too early: $P(.A)$.

Such contents are coded with capital P (Past) and, when possible, the exact past-period is coded between parentheses following the capital letter P.

3.12.2. The subject wants to avoid something, or is afraid that something may not happen or may happen too late.

Examples: *I would regret it very much if...my new car would be in a collision*: code Y-
I fear that...my children will quarrel for their inheritance: code $.X$
I would regret it very much if...I would become pregnant again: code Y- ← p (p because the subject implicitly refers to the past). We assume that the subject has in mind the future moment where the expressed negative motivational object normally happens, and not the whole period during which it could happen.

When there are reasons to assume that the subject considers the total period, we code the whole period.

Examples: *I am afraid that ...I'll never succeed in my examinations* (E);
I would regret it very much if...I would never be able to buy a house (L);
I wouldn't like...to be a bachelor my whole life long ($\underline{L}$).

3.12.3. The subject says that he does not like a certain temporary situation, or something that may last for some time or even forever.

Examples: *It displeases me that...it is raining* (D), *that our house is too small* (present life period of the subject), *that my husband does not understand me* (l). *I am not inclined to...change my character or my way of living* (l). *I think it is sad that...the Third World does not develop faster* (x).

3.12.4. The subject who fears that something may happen too early is assumed to have in mind also the moment when it normally should happen and which is by definition farther in the future (cf. p. 69). We code the most distant temporal localization.

Examples: *I am afraid that...I will not live very long* (O.);
I would not like it if...my son would have children much too early in his life (A_2). The code refers to the father's life period. We code A_2 because that is usually the life period *during which one becomes a grandparent,* although the subject will probably not yet be in A_2 if his son becomes a father much too young.

3.13. *Motivational objects for others*

As shown in the last example, the code refers to a temporal period in the life of the subject completing the MIM and not to a period in the life of the person he is referring to (except for the goals to be situated in the historical future: symbol x).

For example, the sentence completion expressed by a pregnant woman *I hope that my father will still be alive when my child is born* is coded Y. The example given under 3.6.3.:...*that my parents will continue to stay with us* refers to a present situation of which the subject wants the continuation. It will, however, not continue his whole life (subject in A_1). We code $l(A_1)$.

3.14. *Atemporal objects and uncodable responses* (symbol: ?)

Here are two categories of sentence completions that are clearly different from the point of view of motivational content analysis, but neither of them can be coded for time perspective.

3.14.1. Totally unrealistic or fantastical objects are considered to be a-temporal. Example:*I would like to marry the moon, to be Napoleon.*

3.14.2. Sentence completions with no motivational object or that are illegible are uncodable. Also responses such as: *I hope...it is all in vain to hope; I want...nothing;* and *I am doing my best to...because I am obliged to*; are uncodable for time perspective.

This type of responses usually has a very low frequency and is coded '?'. The validity of the responses given by a subject who gives a relatively high number of such answers should be questioned.

However, responses such as *I would like...to be Jesus Christ*, or *...to be a boy rather than a girl*, may express that the subject is unhappy with what he is and that he misses certain personality characteristics. When they are useful for a certain type of research, they should be coded as personality characteristics (code *l*).

3.15. *Global time categories*

After coding the time localization of the goal objects of a group of subjects it may be useful to describe them qualitatively in terms of a few global time categories such as the *immediate future*, the *intermediate future*, and the *distant future*. The immediate future covers the periods T, D, W, M; the intermediate future covers the periods Y, Y- and the current life period of the subject. The distant future contains all the temporal periods beyond the current life period of the subject and all objects coded L (or $\underline{L}$). The 'open-present' and the historical future form two special time categories that have their importance for a qualitative description and interpretation of the data. After these preliminary calculations, several measures of the extension and density of the subject's time perspective have to be made, as described in Chapter 4.

As an illustration of the rules given in the present section, a list of examples follows for each of the temporal code symbols. In difficult cases, the coder will ask himself the question mentioned above: 'Where in time can a subject's mental look be assumed to dwell when expressing this motivational object?'.

As a conclusion, it should be noted that a certain vagueness belongs to the psychological reality of a subject's future time localizations. Our 'measures' should not pretend to be more precise than the psychological reality itself.

IV. LIST OF EXAMPLES

T
    to be through with this test.
    to leave this room.
    that I accepted to participate in this research.
    to help you by answering this test.

D
    that the weather would be nice tomorrow.
    to have a cup of coffee.
    to go dancing tonight.
    to smoke a cigarette.
    to call my parents.
    to sleep well tonight.
    to pick up my children after school.

W
    to go to the football game.
    that the weather would become better.
    to take a day off.
    that my cold would be over.
    to write a letter to...
    to have a haircut.
    to buy a new record.

W̲
    that it would be raining the whole week long.
    that I should have to stay in bed for a few days.
    that this week seems so long.

M
    to lose some weight.
    to take care of my garden.
    to buy a new dress, a bicycle.
    to take a few days off.
    to see my dentist.

M̲
    to enjoy my summer vacation starting tomorrow.
    to do nothing else than to study this month.
    that I have to wait a month before meeting my girlfriend again.

Y
    to spend my vacation in the mountains.
    that the summer will be nice and warm.

Y	to cultivate roses and vegetables.
to prepare my exams.
to save some money.
to make a nice trip.
to paint my house.
to buy some furniture.
to succeed in my exams.
$\underline{Y}$	that I have to stay in this clinic for a whole year.
that my military service takes the next 12 months of my life.
to stay in bed for the whole duration of my pregnancy.
$Y.$	to find a job at the end of this academic year.
to succeed in the exams at the end of this academic year.
to celebrate Christmas with my parents.
$Y\text{-}$	to marry soon.
to do my military service.
to enlarge our summer cottage.
to buy a new car.
to get used to this new country.
E	to enjoy my school years (for a subject in E_2; otherwise E_3).
to continue my studies in a foreign country (idem).
to obtain a very good education.
to buy more books.
to work hard in school.
$E.$	to finish school.
to obtain my final diploma.
the day that I will not have to study anymore.
$\underline{E}$	to work hard as long as I go to school.
that it takes so many years to become a lawyer.
to take a summer job to pay for my studies as long as I have to.
E_0	with P: memories of the first 6 years of life: $P(E_0)$
that my mother died when I was born.
that I had to go to that kindergarten.

E_1 (with P) memories of the primary school period: $P(E_1)$
that my parents forced me to go to the scouts.
to go to Sunday-school.
that I had to do third grade twice.

E_2 to take additional courses in French.
that my friends are more interested in athletics than in their studies.
to become a member of our high school band.
to go to another high school.
not to go to a boarding-school.

E_2.   to obtain my diploma as a plumber.
when I'll not have to study Latin any more.
to finish high school.

E_3 to become a psychologist, lawyer, etc.
to enjoy my freedom in college.
to prepare a career as a scientist.
to take some free courses.
to become a cheerleader.
to do my graduate studies at the University of...

E_3.   to obtain my diploma as an engineer.
when my higher education will be finished.
when campus life will be over.

$E\text{-}$ to prepare myself for adult life.
to find my true personality.
to find a fiancé.
to date many girls.

$E_3 A_1$ to become more realistic (subject in E_3).

$A_0 A_1$ to become a really professional one (subject in E_3 or A_0).
to make useful social contacts.
to learn several languages.
to know more about social and political life in our country.

$E_3 + A_0$ to go to the university or into the army.

EA	to achieve something important (subject in E).
	to realize my potentialities.
	to help my people.
A	to be successful in my career.
	to earn my living.
	the struggle for life.
	to become someone.
	to realize my goals, plans.
	to make a world trip some day (subject in A_1).
	not to become a slave of my job.
$\underline{A}$	never to work for a boss.
	to keep informed on new developments in my job as long as I am professionally active.
	that my work will never be more important than my family life.
	to go out with my wife very frequently ($\underline{A_1}$ if subject is in A_1).
.A	to marry.
	to find a job.
	to establish myself as a dentist.

Note:

For subjects who are at the end of their E_3 or A_0 period or who just started their A period, the beginning of the A (.A) period may be within a two-year period. We then code Y- rather than .A).

A_1	to build (or buy) a house.
	to earn a promotion in my job.
	to adapt to my new job.
	to spend more time with my children and family.
	to improve our family situation.
	to arrange our house nicely.
A_2	to marry off my children.
	to become very famous.
	to become the general director.
	to become a member of the board of trustees.

A2	to win the confidence of my grandchildren. to direct my children to college. that my son will be successful in college later.
AO	to have a happy family life. to have a good relationship with my family. to enlarge the power of our family. to help my wife. to make my children happy.
<u>AO</u>	to stay always with my husband.
A2O	to become rich one day. to take care of my grandchildren. that my children would love to visit their parents.
A2.	to arrange my succession. to prepare my successor.
O	to be alone when I'll be old. that my pension will be sufficient to live on. to be retired. that I'll have more time when I'll retire.
.O	to retire. to move to Florida the first day of my retirement. to find something to be occupied with when I retire.
O.	to die. to have a peaceful death. to work till the last day of my life.
.X	that my children will not quarrel about their inheritance. that my son will not sell our family house.
X	that our children will keep contact with each other. to survive in my paintings, books. that our sons would continue the family business after my death. to go to heaven. to be reunited with my husband in heaven.

Examples 129

OX	that my grandchildren be successful in life (subject in O). that our children continue our family traditions (subject in O).
L	to have a successful life. to fulfill my life duties. that my life will be worthwhile. to defend always the poor, or...
$\underline{L}$	to live my whole life in this city. to be able to work as long as I live. to devote my whole life to... to enjoy the affection and love of my children for the rest of my life.
l	to be popular, attractive, intelligent, happy, good. to be an influential man. to be a good Christian, Communist, colleague, etc. to be respected by my friends. to be able to speak French. to look like a really happy housewife.
$\underline{l}$	to stay in good health. to be always friendly. to keep my faith in God. to always enjoy the affection of my children. to always look happy.
$l(E)$	that my parents would understand me. that professors are not very social. to be able to motivate my friends in school.
$\underline{l}(A)$	to stay honest in my professional life. to always be motivated for my work. to keep enjoying my work.
$l(E)$	to be very motivated as long as I go to school.
x	peace for all people, peace on earth. that the rich world help the poor world more efficiently. a better understanding between generations. a worldwide revolution.

	the Indians may keep their reservations.
$P(...)$	that I did not go to high school. $P(E_2)$ that I did not allow my son to go to college. $P(A)$ that my parents refused to come to my wedding. $P(.A)$ to relive my childhood. $P(E_0E_1)$ to redo my college years. $P(E_3)$ that I went to the movies yesterday. $P(D)$ that I did not succeed in my exams last year. $P(Y)$
$\underline{l} \leftarrow P$	to be always optimistic as I always have been.
$l \leftarrow P(E_0E_1)$	to be as happy as I was in my childhood.
$l \leftarrow P(.A)$	...that I am not a bachelor anymore.
$A_0 \leftarrow P(A_1)$	to drink much less than I used to (subject in A).
$A_2 \leftarrow P(A)$	not to work so hard anymore as I used to (subject in A_2).
$Y \leftarrow P(Y)$	to work harder this year than I did last year. to visit Italy, as I did last year.
$E_3 \leftarrow P(E_2)$	to spend more time studying than I did when I was in high school.
$Y \text{-} \leftarrow P(E_1)$	to meet my childhood friends again.
$M \leftarrow P(W)$	that I invited my neighbours last week for our coming party.
$Y \leftarrow p$	to visit Europe again next year. to go back to Spain for my next summer vacation.
$\underline{L} \leftarrow p$	never again in my whole life...
$L \leftarrow p$	never again...
$E \leftarrow p$	to study more seriously.
?	to be the sun.

? to be able to do miracles.
...because I have to.
totally unrealistic or fantastical objects.
illegible sentence completions.

CHAPTER VII

MANUAL OF MOTIVATIONAL CONTENT ANALYSIS

by J.R. NUTTIN & W. LENS

I. PRINCIPLES OF OUR CONTENT ANALYSIS

II. THE STRUCTURE OF THE CODE

III. THE CODING TECHNIQUE
    1. The subject's personality: the Self (S)
    2. Self-Realization (SR)
    3. Activity and work: Realization (R, R_2, and R_3)
    4. Social motivations: Contact (C, C_2, and C_3)
        4.4. The code for people
    5. Cognitive motivations: Exploration (E)
    6. Transcendental objects (T, Tr, and Tx)
    7. Possessions (P)
    8. Leisure, recreation, and pleasure (L)
    9. The code for modalities
    10. Negative components (n, $n+$, and np)
        10.5. Aggressive tendencies
    11. Responses to negative inducers (-)
    12. Motivations related to the test (Tt)
    13. Unclassifiable and blank responses (U)
Conclusion

IV. LIST OF EXAMPLES

CHAPTER VII

MANUAL OF MOTIVATIONAL CONTENT ANALYSIS

by J.R. NUTTIN and W. LENS

The Motivational Induction Method (MIM) is used to obtain samples of motivational objects as they exist in homogeneous groups of subjects. The method is not assumed to induce the motivational state itself, but rather the *verbal expression* of subjects' motives, concerns, aspirations, and fears. A general description of the method, together with a discussion of its validity and reliability, have been given in Chapter II. The reader is presumed to be familiar with that chapter and with the instructions and list of inducers before studying this *Manual*.

The motivational content analysis as described in the present chapter can be done independently of measuring the subjects' time perspective. It should be noted, however, that motivational content analysis and future time perspective are complementary to a certain extent, since the temporal localization of a motivational object is somehow related to its content. There is a strong relation, indeed, between *what* one wants and the time interval needed to obtain or to realize it.

I. PRINCIPLES OF OUR CONTENT ANALYSIS

The theoretical foundation of our content analysis is situated in the way in which motives and needs can be identified and classified according to their essential components. This problem is discussed in detail in our theoretical volume (Nuttin, 1984). The general aim, then, of this analysis is to classify a subject's specific motivations in terms of the category of objects he is motivated for and the kind of behavioral relation he wants to establish with that object. In fact, the nature of the object involved and the behavioral relationship to *be established with that object define the subject's motives and needs.* Thus, the hunger need is identified by the kind of objects involved (objects fulfilling the function of nourishment, called *food*) and by the behavioral relation of *eating* that kind of object. When I am motivated 'to meet a friend', the object 'a friend' and the behavioral relation 'to meet' him define the nature of my motivation. When I am motivated 'to better understand my friend', the material object is the same ('my friend'), but the behavioral relation is different ('to understand him'). The same holds for motivations such as 'to buy a house' and 'to paint my house', etc. The *global* object of my motivation is 'painting my house' or 'understanding my friend'. The 'material' object and the behavioral relation towards that object together constitute one meaningful behavioral act. This global behavioral object is what defines an act and what is usually meant when we talk about a subject's 'motivational object'; it is the direct aim of an act and its intrinsic end (its immediate or objective goal). Philosophers have called it the *finis operis*, i.e. the end or goal of the act itself, in opposition to the *finis operantis* (the aim of the actor), i.e., the personal purpose the subject has in mind when performing the act. Thus, the subject's personal purpose in painting his house may be to please his wife. A student who says that he wants to go home may actually want to do so to ask his parents for money to buy a car. It is important to note that our content analysis is not concerned with the subject's further aims insofar as such means-end structures are not explicitly mentioned in his MIM responses. It is probable, however, that a subject who wants to go home to ask his parents for money to buy a car will also express his desire to buy a car in one or other of his sentence completions. As a rule, our content analysis is limited to the motivational or behavioral objects explicitly mentioned by the subject.

    Another basic principle of our analysis can be formulated as follows. Beyond the specific object expressed by a subject, we are looking for *the more general function or role* played by that object in the behavioral interaction between an individual and his world. Thus, different types of behavioral activities for getting objects to eat are not coded in terms of the specific objects or behavioral acts but in terms of their more general function: in this case the function of satisfying the subject's need for physical self-preservation. In the same way, aspirations for different kinds of interpersonal relations are coded in terms of a need for social contact, etc. In other words, the more general need underlying the subject's desire for a specific object is coded. To that end, the classification of objects and behavioral relations as discussed in Chapter II is very helpful.

Finally, the symbols used for coding or classifying the motives are composed of different elements representing the main components of a motivational object. These components refer to the 'material' object of a subject's desire and to the kind of behavioral relation implied in that desire (e.g. to know or to understand that object, to get in social contact with it, to possess it, to relax or amuse oneself). A subject who wants to become more generous with his friends is motivated for something related to his own personality (category of the Self); what he wants is of a social nature (generosity); the persons involved are his friends; and a developmental modality is implied in that motivation since he wants to make some progress (to become *more* generous). As will be shown below, all those components can be expressed in the code symbols used. Thus, each motivational object may be coded in terms of several components that are considered of some interest in the frame of a specific research topic. Hence, in looking at the code symbol, the nature and the structure of a motivation can be discovered. It is possible, for instance, to calculate the increase or decrease of a given component or of a specific modality as a function of a differential or experimental variable introduced in a group of subjects.

The coding system as proposed in this chapter is the product of empirical research. Although based on a theoretical concept of motivation, it was tried out by analyzing a great number of individual protocols obtained from all kinds of subjects in different cultural settings. It proved to be adequate for classifying motivational objects in males and females between the age of 12 and 80, literate and illiterate people - the latter group by oral application of the MIM - in Eastern, Western, and primitive cultures. Moreover, the coding system is flexible enough to be adapted, if necessary, to the specific goals of a special research project.

II. THE STRUCTURE OF THE CODE

As explained above, coding the content of motivational objects is done by combining several elements. Before giving a detailed description of the coding procedure, the constituent elements and their combinations have to be briefly discussed. We distinguish four types of elements: first, the symbols for the motivational *main categories*; the different *subcategories* within a main category, which are coded by adding a second symbol to the first; a set of symbols is used to code certain *modalities* characterizing the subject's motivation; the symbols that refer to the category of *persons* that are involved in the motivational object.

The main element in a motivational code is a capital letter referring to one of the motivational main categories. Table 10 lists and describes these categories. Their meaning will become more precise when we further describe the subcategories and additional symbols.

It will be noted that the fourth main category mentioned in Table 10 - labeled *social contact* (C) - is divided into three subgroups (C, C_2 and C_3) according to the three types of social interaction described in Chapter II (cf. p. 50-52). The same is done for the category of productive activities: R refers to activities in general, R_2 to professional work, and R_3 to the 'professional' activities of students, that is studying. For non-student subjects, studying or reading may belong to the category of leisure activities or to the category of exploration, depending on the specific content of the subject's response and its context. In addition to the capital letter - sometimes followed by a number - a complete code may have three types of supplementary symbols:

- first, small or lower-case letters that immediately follow the main symbol. For practical reasons, we use abbreviations (example: $Spre$ for Self-preservation);
- second, a combination of capital and small letters, written between parentheses and following the main symbol (example: $S(Calt)$ for personality characteristics involving altruistic contacts);
- third, graphical symbols (for example an arrow →).

These specifications are used for:

- classifying an object in a subcategory of a motivational main category;
- coding modalities characterizing the motivational object (code for modalities, cf. p. 159);
- coding persons that are referred to in the motivational object (code for people, cf. p. 151).

Not all coded motivational objects have these three elements. Sometimes the goal object mentioned does not have any specification, and only the capital letter for the main motivational category can be coded.

We realize that one's first contact with the coding system and the many combinations that are possible may be rather disconcerting. Human actions and their motivations are, indeed, complex structures . However, the transparency of the coding system and its symbols promotes familiarization.

The examples that will be given when we describe each motivational category and an extensive list of additional examples at the end of the *Manual* will facilitate the application of the coding technique.

III. THE CODING TECHNIQUE

It is necessary, first to read all the responses of an individual before coding their motivational content. Indeed, the context of a particular response may affect its meaning or content. A correct coding sometimes requires taking that context into account.

For the sake of easy manipulation of the data, we ususally transcribe the responses of each subject on two or three sheets of paper on which only the numbers of the inducers are printed. Two columns at the right side of the page are reserved for, respectively, the content code and the time perspective code. We refer also to the general advice given in *Article I* of the *Manual of Time Perspective Analysis* (cf. p. 104).

1. The subject's personality: the Self (S)

The category of Self and the categories for work (R) and social contact (C) require many specifications because of their importance in human motivation.

Two main types of specifications are made within the category *Self*:

1.1. the first is related to motivations with regard to the *global* personality:

S = the *Self* as a whole, i.e. without any specification;

140   Manual: Motivational Content Analysis

> Sc = *self-concept*: the concept or perception one has of oneself;
>
> $Spre$ = *self-preservation*;
>
> $Saut$ = *personal autonomy*.

1.2.   The second type of specification is related to *partial aspects* of the personality:

> Sph = *physical*: motivations related to the physical aspects of the subject;
>
> $Sapt$ = *aptitudes* (intelligence, aptitudes, special capacities, etc.);
>
> $Scar$ = *character* = character and personality traits as far as they are no aptitudes or abilities.

1.3.   The small letters used as symbols for these specifications are added directly to the capital letter S. A second category of elements are added between parentheses to S. They specify the field in which the motivational object is situated. For example, personality characteristics ($Scar$) may have a *social component* that is not expressed by the symbol $Scar$. Coding these components makes it possible to determine afterwards in which goal objects these components were present. The symbols used for coding such motivational components are derived from the symbols for the main motivational categories. For example, the symbol C is used for a *social* component and $R2$ for everything that refers to professional work. Additional letters are used to distinguish the *altruistic* and *egocentric* aspects of social contact. We use the following additional symbols in the category *Self*:

> $Scar(Calt)$: this symbol is used when the subject wants to have an altruistic personality trait (C = social contact; alt = altruism);
>
> $Scar(Ceg)$: egocentric, competitive, dominating personality trait (C = social contact; eg = egocentric).

Note: Because the symbols $Calt$ and Ceg, following the capital letter S, unequivocally refer to personality traits (being altruistic or egocentric), it is unnecessary to add '*car*' to the capital S. The codes become therefore $S(Calt)$ and $S(Ceg)$

TABLE 10

THE 'MAIN CATEGORIES' in MIM-CONTENT ANALYSIS

---

1. *S (S = Self)*: motivational objects related to some aspects of the personality (personality characteristics, etc.)

2. *SR (SR = Self-realization)*: activities or endeavors towards the development (actualization) of the subject himself (self--development)

3. *R (R = Realization)*: all forms of productive activity or 'work' intended to produce objects of some utility (as opposed to leisure activities).
   R = activities in general;
   R_2 = professional activities;
   R_3 = study activities.

4. *C (C = contact)*: motivational objects involving some form of social contact:
   C = contact with others;
   C_2 = when one wants something from someone else (reciprocal contact);
   C_3 = goals formulated for others.

5. *E (E = exploration)*: activities that are related to the desire for information, knowledge, explorations, etc.

6. *T (T = Transcendental)*: motivational goals or values that are religious, existential, or transcendental in nature.

7. *P (P = Possessions)*: desires to possess or to acquire an object.

8. *L (L = Leisure)*: activities that are related to recreation, play, leisure, (as opposed to work).

9. *Tt (Tt = Test)*: responses that refer to the MIM application.

10. *U (U = Unclassified)*: unclassifiable and blank responses.

---

rather than $Scar(Calt)$ and $Scar(Ceg)$.

Examples: ... *to be 'open' to others* = $S(Calt)$;
... *to be the most popular girl in my class* = $S(Ceg)$.

$Sapt(R_2)$: motivational goals about particular aptitudes can be specified by indicating the field in which the desired aptitude applies. When the aptitude is related to professional work we code R_2 between parentheses after the symbol $Sapt$. An aptitude that would be used in recreational activities is coded $Sapt(L)$.

These coded specifications are useful because they make it possible to find in all expressed motivational goals those that are related to, for example, professional activities, leisure, etc. (motivational components). Goals that are coded in terms of R_2 or L are directly related to that field of activities. When these symbols are coded between parentheses, they express a specification or component that belongs to another main category of motivational objects. Thus, a motivational object coded $Sapt(L)$ is related to S and to leisure activities.

1.4. Goal objects that are related to the global personality can also be specified by adding, between parentheses, the field in which the goal is situated. We use abbreviations to make it easy to understand the code.

$Saut(ec)$ = economic autonomy: the subject wants to be economically or financially independent or autonomous. When the person(s) of whom he wants to be independent is (are)mentioned (to be financially independent of my parents), we add the code for that (those) person(s) (cf. p. 151).

$Saut(ph)$ = physical autonomy: elderly people, physically handicapped subjects, or prisoners often express the desire to remain physically independent or the fear of becoming physically dependent on others.

$Saut(ps)$ = psychological or social autonomy is used for independence in general (except for physical and economic autonomy).

$Spre(ec)$ = economic self-preservation: for example, the desire for a stable personal economic situation (...to have enough money to live;... etc.).

Spre(ph) = physical self-preservation:...*to stay healthy;...to recover from an illness;...not to die in a war, or in an accident*, etc.

Spre(ps) = psychological self-preservation: to defend and to protect one's equilibrium, stability, identity, psychological well-being. Somewhat neurotic people are afraid of anxieties and agonies and want to be calm, relaxed. They want to defend themselves against criticism and denigration by others, or they fear being disappointed in their goals and aspirations.

Spre(T) = self-preservation on the existential or transcendental level: existential goals concerning death, nothingness, eternal life; the desire to survive in one's children, books, etc.

2. Self-Realization (SR)

The symbol R is used for activities and realizations in general. However, when the subject's activity is directed to developing his own personality and potentialities, the codes S and R are combined in the symbol SR for self-realization. It is used when the object of the individual's activity is the development or realization of his personality, some aspects of it, or his potentialities. Each motivational goal that can be coded S is coded SR when the subject expresses an activity or an effort he is motivated to do to attain that object. This category of motivational objects is related to the general need for self-realization or self-development.

All combinations of symbols mentioned for the category S can also occur in the field of SR, except for the code $Spre$ (self-preservation). Self-preservation or self-defense can hardly be considered as a way of self-development.

The following rules will be helpful in deciding when to use the code symbol SR.

2.1. A fundamental condition for coding SR is that the subject considers his global personality and that he refers to a certain activity or development. The subject wants *to become* (or not to become) a certain kind of person, or he wants *to lead* a certain way of life. Such goals are coded SR. The fact that the subject refers to his life or his 'being a certain person' is sufficient to assume that he considers his global personality. Verbs such as *to become, to avoid*, or these referring to some other forms of activity imply a certain type of 'realization'. The same holds for expressions referring to progress or development (to become *more*...). Even the expression *to stay* (for example, to stay honest) means that the individual is motivated to not give in to opposite tendencies.

Examples: responses such as *to lead a honest way of life, to become a honest person*, and *to live as a honest woman*, are coded $SR(C alt)$. But the response *to be honest* is coded S (without R). Honesty is a social personality characteristic that benefits others (therefore: $C alt$ = altruism). The verb "to be", in this example, does not explicitly refer to the global personality or life. Hence, the subject wants to have a personality characteristic, rather than 'to realize' a specific kind of personality.

2.2. The symbol SR is also used when the goal refers to a special aspect of the personality and when the subject considers that aspect as a specific way of personality development for himself as a whole. For example, the subject wants to become *Miss America, a medical doctor, a sailor* or *carpenter*. Someone else intends to *fully develop her intellectual qualities*. In all these examples, the subject has an active aspiration for self-development or self-realization. When the self-realization must be situated within the professional field, we use the symbol R_2 (or R_3 for studies) instead of R, and we code SR_2 or SR_3. The person who wants to become *a scientist* or *a medical doctor, a carpenter*, etc. refers to his or her professional life as a field for self-realization. We code SR_2 (and not SR). In the example *to become a scientist* we can even specify the professional component by adding between parentheses the symbol E (exploration) to the code: $SR_2(E)$.

2.3. Sentence completions in which the subject talks about realizing his projects, plans, ideas, to succeed in life, in a career, in enterprises etc. are understood as expressions of a desire for self-realization. Indeed, the global personality is identified with these projects. However, we do not code SR for a more particular or limited enterprise, plan, or success (in the singular). The individual who is striving to succeed in his exams has a much more limited project in mind than the subject who strives for success in life in general. This difference is also important for measuring time perspective (cf. p. 103).

Note: When analyzing the motivational content of certain sentence completions, we are directly confronted with the great variety and flexibility of verbal expressions (cf. p. 50). It was, therefore, necessary to formulate some rather arbitrary rules and distinctions. For example, the difference between the expressions *to be honest, to be a honest person, to stay honest, to become more honest,* etc. is open to discussion. We made it a general rule that motivations expressed in terms of *to be* are assumed to reveal less effort and engagement than those expressed in terms of *to become, to stay, to remain, to lead a life,* etc. Therefore, the latter are automatically coded *SR*, whereas motivations in terms of *to be*, are classified in the *S*-category; that is also the case for motivations such as *to be more honest*, as opposed to *become more honest*. The use of certain words is here - more or less arbitrarily - our criterion.

3. Activity and work: Realization (R)

The category *R* refers to the domain of activity and *work*. It includes the great variety of activities performed on objects in order to obtain some benefit from it (in opposition to recreational activities).

3.1. Two important subcategories are distinguished in the main category *R* in addition to the *R*-category without specifications:
   R 2 for professional activities;
   R 3 for full-time study activities, preparing for productive adult life.
   R 2r is sometimes used for activities related to a second job.

3.2. A small letter 'a' may be added to the symbols *R, R 2*, and *R 3* when *achievement motivation* is involved. When the subject is concerned with a *level of excellence* or *success* and when he expresses *some kind of competition* or, in general, a *level of performance* in terms of a norm of excellence, achievement motivation can be coded. We propose using this specification also for all motivational goals which by their very nature refer to a high level of aspiration, a spirit of competition or an explicit desire for success. For example, the student who works for a 'summa cum laude' in his exams gets a code *R 3a*; the engineer who tries to invent a new technology gets a code *R 2a*.

However, the intentions such as *to work hard* or *to be persistent in his work* are not considered as expressions of achievement motivation. A certain level of performance or norm of excellence is required. Motivational specifications such as maximal effort are coded as *modalities* (cf. p. 159).

3.3. A special kind of motivational object in this area is related to rest and refusal to work. We distinguish two types:

nR : the intention or plan not to work: *I don't want to study* is coded *nR3*. A small letter *n* preceding the code means that the subject does *not* want to work or to do specific work (*nR2*). This small *n* is also used with other motivational categories:*nC* means that the subject does not want social contact, etc.
Aspirations not to work, not to meet someone, etc. are understood as desires for rest, to be alone, etc. *nR* refers to non-activity in the form of rest or refusal to work (see below, section 10).

nR(ph): the symbol small *ph* is added between parentheses when the subject's goal is to have physical rest (for example: *I will be glad when I'll be able to have a good long sleep*).

nR(ps): *ps* between parentheses refers to psychological rest, peace, quiescence.

3.4. Finally, the symbol *R* can be used in combination with the codes for other main motivational categories. For example, the response *to become financially independent* is coded *Saut(ec)*, but the response *to defend my financial independence* or responses in which some type of activity to maintain or to attain that same goal is expressed, are coded *Saut(ec)/R*. The symbol *R* is added to the main part of the code. The two parts are separated by an oblique line. Thus, the symbol */R* can be added to most symbols of other motivational categories. It only means that the subject expresses some kind of activity in relationship to the motivational object. However when the activity has to do with the development of the subject's personality, the symbol *SR* is coded as the main category.

4. Social Motivations: Contact (C)

An important group of human motivations is related to behavior towards other people. Our term *contact* is meant to refer to all these kinds of behavioral interaction that may happen between people. The specific category of people implied in such social contacts often determines their motivational nature (erotic, parental, intimate, superficial, etc.). Several categories of people are taken into account by our coding technique, so that a thorough analysis of social contact becomes possible. The special *code for people* is discussed in paragraph 4.4.

Within the broad category of social motivations, three subcategories have been distinguished, as mentioned already earlier:

C refers to motivations to get in one or other form of contact (behavioral interaction) with one or more people. The negative form if this motivation is to avoid such contacts or to stay alone.

C_2 refers to motivations for reciprocity: The subject wants *to be contacted* in some way *by others.*

C_3 refers to motivations in which a subject expresses his wishes, aspirations, or fears *for others*: He wishes, for instance, his father to find work, to recover, etc. Most objects of motivation for oneself can be also objects of wishes and fears for others.

Within each of these three subcategories, several types of social contact are to be distinguished, besides the differentiation that is introduced by the kind of people involved.

4.1. *To get in contact with others (C)*

In the category of social contact, the symbol C refers to a motivation that is often specified by the kind of person that is involved. For example, a subject's desire *to go out with a girl* is assumed to be a motivation for a more or less erotic contact (cf. p. 152). The motivational intention *to write a letter to my parents* implies a quite different type of social contact.

4.1.1. Three kinds of social contact are coded by adding lower case letters to the capital letter C:

C_{int}: *intimate contact.* For example, the goal object 'to marry' is coded as a desire for intimate social contact (C_{int}) with the specification 'erotic' (symbol e). The total coding being then $C_{int,e}$. The desire *to have a good friend to whom I can entrust something* is a wish for a type of intimate social contact (C_{int}), but in the field of friendship (small letter a). The total coding being $C_{int,a}$.

C_{alt}: *altruistic contact,* i.e. for the benefit of the other person. When the subject not only wants altruistic contact, but also wants to do something in an active way for the benefit of the other, *Calt* is followed by /R. For example, the goal *to help my friend* is coded *Calt,a/R*. The *activity* is coded by the capital letter R following the main code and separated from it by an oblique line. The *friend* the subject wants to contact in an altruistic way is coded by the small letter *a* (preceded by a comma).

Ceg: *egocentric social motivations*. The term *egocentric* is taken in a very broad sense, it does not necessarily imply egoistic tendencies. An individual who wants to *convince* someone else, or who wants to *influence* or *lead* other persons, wants to establish an *egocentric* social contact in the sense that he places himself in the foreground in the social relation. *To do all I can to convince my friend that I am right* is coded *Ceg,a/R*. The same code is used for sentence completions in which the subject wants *to put much effort in showing his friend what he is able to do*. It will be noticed that the specific content of the motivation is lost in the code, but its egocentric nature and the specific person(s) involved are reflected in the symbols.

4.1.2. The symbols *Calt* and *Ceg* can be added between parentheses to the code *Scar* (personality trait) to indicate the *social* (altruistic or egocentric) nature of the motivational object. For example, *to be more helpful* is coded *Scar(Calt)*, the first part referring to a personality trait and the second part specifying its social-altruistic nature. The category of person to whom one wishes to be more helpful can be added, as will be explained later: *to be more helpful for my friend*: *Scar(Calt,a)*. This way of coding the different components of a motivational object makes it possible to analyze them separately. For example, we can calculate the relative frequency of the social component (i.e. of the symbol *C*) as an additional component in the motivational objects of a group of subjects and not just its frequency as a main category. The same holds for each motivational component.

4.1.3. A desire for 'existential' or 'transcendental' contact (cf. p. 156) is manifested in some religious, pantheistic, or mystic tendencies. In such cases, the symbol for the 'transcendental' domain (*Tr* or *Tx*) is added between parentheses as a specification to the symbol C. Aspirations *to be in contact with a supreme Being* or *to feel united with God* are coded respectively $C(Tx)$ and $C(Tr)$.

4.1.4. Negative types of social contacts, such as the desire *to be alone*, or *to do harm to others*, are discussed later (see p. 162). Other modalities of social contact (for example, the desire *to improve* one's social contacts) will be discussed in the *Code for Modalities*, cf. p. 159.

4.2. *Social reciprocity*: wanting something *from* someone else (C_2)

The symbol C_2 is used for motivational objects in which the subject wants (or does not want) *to receive something from others* or that others *would do* (or *would not do*) *something to him*.
The others play the active role and the subject himself is the object of their actions: he wants (or avoids) those actions. The first element of the code is C_2 to indicate that the subject wants something from someone else. The second element of the code identifies the type of person *from whom* something is wanted (see Code for people, p. 151). The third component is coded between parentheses and refers to *what* is wanted or expected from the other(s).

4.2.1. Three categories of social contact can be distinguished among the positive things that the subjects may expect from others:

- affection (af);
- appreciation (ap);
- support (su).

$C_2(af)$: the subject expects from someone else feelings of sympathy, love, forgiveness, understanding, comfort, etc.

Examples:...*that my friends would show me their sympathy* = $C_2,a(af)$;
...*that my wife would love me* = $C_2,e(af)$.

$C_2(ap)$: the subjects wants to be esteemed, appreciated, respected, recognized for his merits or capacities, admired, approved, accepted in the group, etc. The distinctive feature is that this type of social relation implies an *evaluation*.

Examples: ...*that my qualities be recognized* = $C_2(ap)$;
...*that my parents* (= p) *would appreciate me* = $C_2,p(ap)$.

$C_2(su)$: the subject wants to receive some type of support or help (most often psychological) to reach a more or less specific goal. He expects encouragement, guidance, cooperation. When he wants to receive some material goods that are not just a help or a means, we code P for possessions, i.e. some material property, for instance: money (cf. p. 157). The domain in which one wants to be helped is coded by using the codes for the main motivational categories. For example, R_2 when the subject wants some help in his professional life.

Examples: *I am sorry that my father does not help me with my job* = $-C_2,p(su,R_2)$;
I am sorry that my father is not willing to pay for my studies = $-C_2,p(su,R_3)$.

The minus sign preceding the code is used for responses given to negative inducers.

4.2.2. There are, however, other categories of social interactions an individual may want (or not want) from others. They are coded with the symbols corresponding to the category of objects that are involved.

For example: *It displeases me that my friends boast in front of me.* 'To boast' is an egocentric type of social interaction (code C_{eg}). The symbol C_{eg} is added to the code C_2,a, expressing that the subject wants something from someone else (C_2), in this case his friend (a). The complete code for this example reads: $-nC_2a(C_{eg})$. The small letter n preceding C_2 is used when the object coded is NOT wanted by the individual.

A second example: *I want my children to be much more independent of me.* In this example, a father wants something from his children (C_2,p) that belongs to the category $Saut$ (cf. p. 142). The code for the example is: $C_2,p(Saut)$. Two more examples:
...*that my parents would offer me a nice vacation* = $C_2,p(L)$;
...*if my father would give me a car* = $C_2,p(P)$.

4.3. Wishes and fears for other persons (C_3)

A subject may wish (or fear) that someone else would (not) do something or have this or that quality or material object or that something would (not) happen to some other person. All categories of motivational objects that subjects want for themselves can be found in the C_3 category as well.

4.3.1. The general rule for coding objects in category C_3 is rather simple: after the C_3 symbol, followed by a comma and the symbol referring to the other person, we code between parentheses the symbol for the motivational object wished or feared for others.

Examples : *...that my father will stay healthy* = $C_3,p(Spre,pb)$;
:*...that my friend will succeed in his exams* = $C_3,a(R_3)$;
:*...that my wife would get more help from her parents* = $C_3,e(su)$.

In the last example, the subject wants another person (his wife) to be helped (respected, etc.) by someone else. In such cases, we code the corresponding code (*su*, *af*, *ap*, as mentioned in paragraph 4.2.1.), but without C_2. The type of person from whom support, affection, or appreciation is wanted can be added to the code. For this example $C_3,e(su,p)$. But coding the two categories of people (here *e* and *p*) is only useful when we are interested in the network of social relations existing in a group of people.

4.3.2. A special category of motivations expressed 'for others' are of a general *humanitarian* nature (= H):

C_3,H : these objects are related to humanity in general (peace, equality among men) or to some minority groups (more justice for the poor; the emancipation of some discriminated groups, etc.). Among negative motivations, we think of the fear for a nuclear war or some other disaster (e.g. an earthquake). These humanitarian motivational objects are to be distinguished from altruistic motivations because they go beyond the possibilities of the individual. They refer not only to something the subject wants to do for others, but also to more global events. The same humanitarian objects may motivate the subject's personal activities in the sense that he or she may engage in some action to attain those goals.

The symbol for humanitarian goals is a capital H (Humanity), and takes the place of the code for people (= C_3,H).

4.4. *The code for people*

4.4.1. We have already referred several times to the code for people who are involved in social motivational objects (types of social contact). The kind of motivation or of behavioral interaction involved is strongly affected by the type of people that are implied in the contact. As such, the code for people refers to specific categories of social relations. The following categories have been generally used in our coding system. The code can be adapted for special research topics.

e (*erotic*): refers to social contacts with people of the opposite sex and that are more or less erotic, such as between boy-girl and husband-wife and all hetero- or homosexual relationships. The social contact between a boy who writes a letter to his girlfriend or who wants to meet her is coded C,e. When the social contact involves a physically or psychologically intimate relationship, we code $Cint$ rather than simply C. Sexual relations and motivations are classified in a different category (see section 8.2.);

f (*family*): refers to relationships between the members of a family (brothers-sisters; parents-children; father-mother if distinguished from husband-wife contacts). Also included are several contacts with members of the extended family (grandparents, uncles, aunts, etc.). It may be useful in some studies to distinguish parent-child and child-parent relationships on the one hand and all other types of family-relationships on the other. We then use the following symbol for the former category of social contacts;

p (*parental*): this symbol is reserved for contact between parents and children, mother and child, etc. The symbol *f* refers to all other categories of family relationships when the code *p* is used;

a (*amity*): refers to social relations with friends of the same sex or contacts with friends in general (unspecified). For friends of the opposite sex, we use the symbol *e* (see above);

gr (*group*): refers to relations with a group as a group. The group can be a school, a sportsclub, neighbors, clients, inhabitants of the same city or village, etc. In some studies a particular group may have a special importance; for example, 'my people', 'the people of my country', the inmates of a clinic, and a prison. We propose the following symbols:

pl (*people*): refers to larger groups to which the subject belongs such as 'my people', 'my country';

in (inmates): refers to the residents of a community such as a prison, military barracks, a school, etc.;

g (general): refers to people in general, everybody (someone, unidentified individuals, etc.);

so (society): refers to the society as an organized entity, or to some type of society (consumption society, etc.);

x (unknown): refers to unknown people, strangers, enemies, etc. unless the goal of the study requires a special symbol for some categories of such people.

4.4.2. Some special types of relations that can be distinguished are:

si (superior-inferior): relationships between superiors and inferiors. If the direction of the relationship is important, one can make a distinction between *si* (superior →inferior) and *is* (inferior →superior);

y (youth): for contact with youngsters, teenagers. Elderly people as a group can be coded *ol* (old age). Conflict of generations within a family can be coded *fy* (family and youth), or *fol* (family and old age);

sp (superior): for special relations with eminent personalities (stars, a president, a very much admired person, etc.).

4.4.3. Summarizing, the code for social contacts has three elements: 1. the symbols C, C_2, C_3, to indicate the general nature of the social relationship (*towards* the other, *from* the other, *for* the other); 2. a symbol to indicate the persons involved; 3. a specification of the type of contact (altruistic, egocentric, intimate), of the social object (affection, appreciation, support), and of the object wanted for others. The symbols of the code can be adapted in function of the specific research goals. Note that this part of our code allows for a thorough analysis of the social network involved in the subjects' motivations.

5. Cognitive motivations and exploration (E = Exploration)

5.1. Cognitive motivations such as the need to be informed or to understand, to explore all kind of objects and situations, and to know the world and oneself are coded E, usually with a specification for the different subcategories.

E : refers to knowledge or curiosity in general:
...*to be informed about many things* = E ;
...*to know more* = E, with a specification for the modality 'development' that is involved in 'more' =$E(d)$. See p. 159 for the code for modalities.

E(S): this code is a combination of 'exploration' and 'self' and is used for self-knowledge:
...*to know myself* = $E(S)$;
...*to know what I can and what I cannot do* = $E(S)$.

If necessary, the specific personality aspect the subject wants to know can be coded by using the symbols for the subcategories within the main category S (for instance: $Scar$, $Sapt$, etc.)
...*to know my aptitudes* = $E(Sapt)$.

5.2. Negative motivations in this category mean that the subject does NOT want to know something about himself:
...*not to be confronted with my true personality* = $nE(S)$.
These negative modalities are explained on p. 162-163.

5.3. When the knowledge is limited to a specific field (for example, professional life or social life) the domain is coded between parentheses immediately following the symbol E. For example: $E(R2)$....*to learn more about new techniques in my profession* = $E(R2)$;

E(C): the desire to understand or to know more about another person is coded with the symbol for social contact (C); if possible, the category of people is also coded:
...*to understand better the reactions of my girl friend* = $E(C,e)$;
...*to understand my children* = $E(C,p)$;
...*a profound knowledge of men* = $E(C,g)$.

5.4. Other objects many individuals want to know more about are:
- life in general (life experiences);
- the social and physical world;
- existential, religious, and transcendental realities and problems.

The symbol $E(T)$ is used for knowledge of existential, philosophical, religious, and transcendental realities and problems in general. The symbol T (for transcendental objects) with its two subcategories Tx and Tr (cf. p. 156) is added to the main symbol E:

...to find the meaning of life = $E(Tx)$;
...to find the truth = $E(Tx)$;
...to find a satisfying answer for my religious questions = $E(Tr)$.

5.5. The desire to know more about life, to have more life experience is coded with the symbol *l* following E (between parentheses) = $E(l)$. It will be recalled that the small letter *l* in the temporal code is used for the category 'open-present', but the context of each code excludes any ambiguity or confusion.

The desire to know more about life or to experience life in all its aspects is sometimes expressed in temporal terms. For example, the individual is eager to be older, to be already in the next life period, etc., but he does not specify in which aspect of life he is particularly interested in. In such cases the same symbol (*l*) is used.
...to enter adult life = $E(l)$;
...to experience the real life = $E(l)$;
...to be a few years older = $E(l)$.

For the last example (to be older; to be further in life) the context of such responses should make it clear if a cognitive need for knowledge or experience is involved, rather than the desire that a difficult, boring, or unattractive time period would end. For example, prisoners, soldiers, students preparing their exams or sick people may express the desire to be a few weeks, months, or even years, older. The subject wants negative objects (a difficulty, an unhappy situation, etc.) to end. In such cases, we code the motivational object (e.g. captivity, professional activity, illness, etc.) with the addition of the modality code for development (*d*) because of dissatisfaction with the present situation (symbol *du*: cf. p. 160).

Example: *... to be one month further in the academic year* (in order to be through with the examinations) = $R3(du)$.
The code simply expresses that the goal object belongs to the study domain ($R3$), that the subject has a negative attitude towards it (*u*), and that he wants it to be already over (*d*).
In the same sense, the desire *to be a few weeks older*, may (from the context) really mean *to be healthy again* ($Spre,pb(du)$).

5.6. Knowledge about natural and cultural objects is coded with a small letter *w* (world) added to the capital letter E.

$E(w)$: knowledge about the social, economic, cultural, political, or physical world (the beauty and mysteries of nature, biological and physical laws); knowledge of other peoples and cultures, their history, their country, their customs. The same code is used for knowledge about artistic creations and scientific discoveries.

The small number of such goals does not justify a separate category. See also the category L (leisure) for some types of cultural goals.

...*to be well informed about oriental civilisations*: $E(w)$;
...*to study tropical botany*: $E(w)$;
...*to make a world trip* : $E(w)$. But when this trip is undertaken for recreation, the code should rather be $L(Ew)$ (cf. p. 158).

Exploration or the search for knowledge in general is sometimes related to the individual's professional life. Scientific research as a professional activity is coded $R_2(E)$. This code must be clearly distinguished from the code $E(R_2)$ which refers to a desire for knowledge or experience about one's professional field.

6. Transcendental objects (T)

Although their frequency is usually rather low, a special motivational category is reserved for motivational objects in the metaphysical or transcendental (T) domain. They are related to philosophical, religious, or existential problems.

6.1. We distinguish the following two subcategories:

Tr: refers to goals in the religious domain (r = religion);

Tx: refers to the philosophical and existential domain (x = existential):

Examples: ...*to believe in God* = Tr;
...*to go to heaven* = Tr;
...*that life is not meaningless* = Tx;
...*to preserve human dignity* = Tx.

6.2. Also negativistic and nihilistic motivations such as the desire *not to exist* or to hope that *life is meaningless* belong to this motivational category. They are coded as negative values ($n+$) (cf. p. 163). This negative symbol follows the symbol Tx or Tr between parentheses:

It displeases me that I was born = $-Tx(n+)$;
My intense hope is that God is dead = $Tr(n+)$.

6.3. The motivational category T can be combined with other categories (cf. supra p.143; 155). An individual may want to realize himself (SR) in propagating a religion:
> ...*to devote my life to missionary work* = $SR(Tr)$.

The subject may want some religious help from someone else or he may have goals for others (C_3) in this domain:
> ...*that my wife would help me in finding God* = $C_2,e(su,Tr)$;
> ...*that my father would find life meaningful* = $C_3,p(Tx)$.

7. Possessions (P)

Acquiring, buying, possessing property is a special way of dealing with objects and, therefore, its dynamic aspect or motivation has a certain identity (acquisitive motivations). Such goals are coded with the symbol P.

7.1. As always, the specific nature of the object that a subject wants to possess or to acquire is coded with the symbols referring to other motivational components (added between parentheses to the symbol P).

P: possession in general (also comfort and conveniences in general):
> ...*to buy (build) a house, a new car*, etc. = P.

$P(Spb)$: for all objects related to the external or physical aspect of the person (clothing, cosmetics, etc.). See p. 140 for the code Spb.

$P(SR_2)$: for objects that have to do with professional activities, and $P(R_3)$ for objects related to study activities.
> ...*to have a new exposition hall* = $P(R_2)$;
> ...*to buy study books* = $P(R_3)$.

$P(L)$: for objects or articles that are used for recreational activities:
> ... *to buy a new record player* = $P(L)$.

The possession of objects used for sensorial or sensual pleasure is coded $P(Lss)$ (cf. p. 159).

$P(Ew)$: articles that are related to cultural activities or cultural education, knowledge of the world or of peoples (cf. p.155):
> ...*to decorate my house with a piece of art* = $P(Ew)$.

$P(an)$: this symbol is used for the possession of animals (an = animal). It can be specified if the animal is needed for professional reasons = $P(an, R_2)$, as a companion = $P(an, C)$, or for the fun of it = $P(an, L)$. Such specifications are taken into consideration when measuring the relative importance of the different motivational components (professional, social, recreational, etc.).

7.2. Acquiring objects for immediate consumption such as food, drink, or cigarettes is not coded P. The subject simply wants to eat or to drink ($Spre$) or to smoke a cigarette (Lss; cf. p. 159). However, when he wants to have substantial quantities of consumption goods (a wine cellar, boxes of cigars) we code $P(Lss)$. When a storekeeper wants to have large quantities of such articles for professional reasons (to sell them) we code $P(R_2)$.

8. Leisure activities, recreation, pleasure (L = leisure)

A large variety of human activities have no utilitarian purposes such as professional services, the production of useful objects or knowledge, but are just done for fun, for recreation, for amusement, for relaxation, and sensorial pleasure.

8.1. What some people do as part of their job may be recreational activities for others; for example, cultivating vegetables or reading and studying books. The context of the responses and the knowledge of the subject's job will allow the distinction to be made.

8.2. Some categories of recreational goals are:

L: relaxation or pleasure in general: vacations, amusement, to have fun, etc.

$L(E)$: leisure activities that are at the same time motivated by cognitive curiosity or the desire for a cultural education (sometimes coded $L(Ew)$).

$L(ph)$: physical recreation, sports, hobbies requiring physical skills (doing things with one's hands such as gardening). When competition or a need to succeed is expressed, we add the symbol for need for achievement (Ra, cf. p. 145). The complete code for competitive sport is hence $L(ph, Ra)$. Passive participation in sports events (to watch a football game, etc.) is simply coded L.

$L(ss)$: refers to all activities for which sensorial or sensual (ss) pleasure is the main reason: eating and drinking for the pleasure of it as distinguished from the need of self-preservation ($Spre(pb)$); sensual pleasure, the pleasure given by smoking, drugs, sunshine, or beautiful weather, etc.

Note: the same symbol can be added (between parentheses) to other motivational components to refer to the sensorial or sensual pleasure that comes together with them. For example:
...to visit a prostitute = $C,e(Lss)$.

8.3. A special kind of recreation or play involves the factor of *chance* or *luck*:

$L(gf)$: this code is used for games and activities in which the chance factor is the exciting and the relaxing component (gf = good fortune). The same factor is also discussed in the code for modalities (cf. p. 161);
...to give it a chance in a casino = $L(gf)$;
...to gamble = $L(gf)$.
When the desire to win money is explicit, the symbol P is added: $L(gf,P)$. When the competitive character of the game is expressed we code $L(gf,Ra)$.

8.4. *Recreation* in terms of rest, not working, doing nothing is coded $L(nR)$: *...to enjoy an absolute rest* = $L(nR)$.

8.5. Obviously other combinations are possible: for example, the motivation to relax by having a conversation or social contact is coded $L(C)$. The social component (C) will then be taken into account when calculating and evaluating the total importance of the contact-component in the subject's motivational goals.

9. The code for modalities

In the same sense as the code for persons specifies the nature of social contacts (C), the code for modalities specifies some formal aspects of motivational objects. Modalities are coded by small letters that follow the content code between parentheses.

We usually code the following modalities in our own research. But, again, this list can be shortened or enlarged according to the aim of the study.

m (maximum): This modality refers to a high level of aspiration (in the broadest sense possible). It is coded when the individual uses terms such as *completely, as much (or as good) as possible, the most* (and other adjectives in superlative degree):

...*to study very hard* = $R\,3(m)$;

...*to develop all my abilities as much as possible* = $SR\,apt(m)$.

In coding the maximum-modality we only take into consideration the content of the sentence completion and *not* the degree of intensity that is expressed in some inducers (for example: *I intensely desire...*).

s (satisfaction): The individual is satisfied with himself or with his present situation. This modality may be expressed in verbs such as *to continue to, to remain, to keep*, etc.;

...*to stay the way I am* = $SR(s)$;

...*to keep this job* = $R\,2(s)$;

...*to continue to be a member of (such and such) a group* = $C,gr(s)$.

d (development): When the individual wants to make progress or to develop or grow in a motivational domain, we code (d). Expressions such as *to become more, to be more, to have more*, etc. manifest a desire for such progress or development:

...*to become more courageous* = $S\,car(d)$;

...*to have more money* = $P(d)$.

The modality 'development' requires that the motivational object is already present to a certain extent. When something new is aspired for (for example: *to become a father*) this modality does not apply.

du (development because the subject is unsatisfied): a progress or development is wanted for an unsatisfying situation or state of being:

...*to find a better job than I have now* = $R\,2(du)$;

...*to reform my egoistic personality* = $S\,car(du)$.

The modality of dissatisfaction is especially interesting when it is expressed in response to positive inducers. The discontent that is implied in the negative inducers themselves is, of course, not coded.

A special case of the modality du was mentioned when we discussed the code $E(l)$ for life experiences (see p. 155).

b (barrier, obstacle): When expressing the object of his desires or fears the individual may also mention that it will be difficult to obtain or to avoid that object. He may say that he expects, or fears, that he will not succeed, that it will not happen again, that he will have too many other things to do, that he will be too busy, that some people or circumstances will prevent or forbid it, etc. Obviously the spontaneous expression of *barriers* or *obstacles* in pursuing motivational objects is often a relevant feature. It may be characteristic for some people or situations:

...*to be myself in spite of pressure from my colleagues* = $S(b)$;

...*to realize my goals notwithstanding the difficulties that I can foresee already* = $SR(b)$.

↔ *(conflict)*: The symbol for conflict is used when the subject expresses two motivational objects or tendencies that are in conflict with each other. Both motivational objects are coded and the double arrow is placed between the two codes. The main difference with the modality *barrier* or *obstacle* consists in that the conflict arises from two objects that are both strived for or feared by the subject and not from *external* obstacles:

Examples: *I want to go dancing tonight, but I should also finish a paper* = $L(C) \leftrightarrow R_3$;

I am afraid that my responsabilities as a father will make it impossible to continue my education = $-Calt,p/R \leftrightarrow R_3$.

When it was decided to code not more than one motivational object for each inducer, we suggest coding the goal that gives difficulties and to add the symbol for obstacle (b). The code for the two examples given above is then $L(C,b)$ and $R_3(b)$.

→ : An arrow is coded for a means-end relationship between two motivational objects (see also p. 118):

I intend to quit smoking in order to stay healthy = $nL(ss) \rightarrow Spre,pb(s)$.

gf (good fortune): a very special motivational modality is to count on good luck, which very often goes together with a passive attitude towards the goal object. While most goal objects are actively pursued by most people, it may happen that someone counts on good luck to obtain a goal. Rather than working and earning money, he participates in lotteries, etc. This motivational modality is coded with the symbol *gf*; it appears to be characteristic for some people.

...*I hope to be lucky in life* = $S(gf)$;
...*to have good luck in my exams* = $R3(gf)$;
...*to have a winning lottery ticket* = $P(gf)$;
...*to win a lot on horse races in order to pay for my studies* = $P(gf) \rightarrow R3$.

Chance and luck play an important role in all kind of games (cf. supra, p. 159). The code $P(gf)$ is used when the subject is more interested in large earnings than in competing or playing. In games where aptitudes or skills are more important than luck, we do not code $L(gf)$ but simply L or $L(Ra)$ when competition is involved (cf. p. 158).

10. Negative components or aspects of motivational objects (n; n+; np)

We use the small letter *n* to refer to negative elements or components in motivational objects. It can be used in three different meanings:

1. the subject does not want to have, to be, or to do something;
2. he wants something that is generally negatively evaluated (to be odious) or he fears (following a negative inducer) something that is positively evaluated (to stay healthy);
3. he expresses a positive goal object in a negative way (I want... not to fail in my exams).

10.1 . The goal is: not to do, not to have, or not to be something (symbol *n*)

The subject simply wants *not to do, not to have, or not to be* something: he wants *not to work,...not to meet his wife; ...not to buy a car*, etc. In such cases, the lower-case *n* precedes the code for the motivational object expressed:

...*not to work* = $nR2$;
...*not to meet my ex-wife again* = nC,e;
...*not to buy a car* = nP.

We code nR (instead of $nR2$) for the refusal of all activity, or the desire to rest. When it is obvious from the context that a rest is wanted to regain a psychological or physical equilibrium, we code

$nR(Spre,ps)$ or $nR(Spre,pb)$. See also p. 143. The same way of coding is applied when such motivational goals are expressed in responses to negative inducers:

I don't want to work in a factory $= -nR_2$.

The minus sign preceding a code does not affect the meaning of the code itself. It simply indicates that the response was expressed to a negative inducer.

10.2. Desires for negative values (n+)

The symbol $n+$ is coded when an object that is generally avoided or considered as having a negative (n) value is positively (+) wanted: something negative is positively desired ($n+$). This type of response is exceptional but it may be very meaningful. In the example *I try to make myself hateful*, we code the personality trait "hateful" ($Scar$), which is social in nature (C) and which generally has a negative value: $Scar(C,n+)$. The response *to be hateful to my friend* is coded $Scar(C,a,n+)$. The desire *I hope that my wife will despise me* is coded $C_2,e(ap,n+)$. The goal object belongs to the category *appreciation* expected from another person $= C_2(ap)$; the small letter e refers to that person (the subject's wife); the symbol $n+$ is used in the same sense for responses to negative inducers:

I try to avoid meeting good, decent people $= -nC,g(n+)$.

10.3. Nihilistic or negativistic tendencies are also considered as having a negative value ($n+$). Due to their existential or metaphysical nature, they are coded Tx or Tr (cf. supra p. 156):

I hope that the world will perish soon $= Tx(n+)$.

10.4. The third application of the negative symbol n concerns rather the form than the content of the expressed goal object. The subject *expresses* a positive motivational object (p) but in a negative (n) way. This gives the symbol np.

In stead of saying *I hope to succeed in my exams* or *to find a job soon*, some subjects formulate their positive desires in a negative way:

I hope...not to fail in the exams $= R_3(np)$;
..........not to stay out of work $= R_2(np)$;
..........not to be without money $= P(np)$.

It is assumed in some motivational theories that the fear for the absence or lack of a desired object is more fundamental than the positive tendency towards that object. Therefore, it may be interesting for some research purposes to find out in what conditions this modality np is manifested and by what type of subjects. As always, the main symbol of the code refers to the object that is positively desired. The modality np does not affect the content of the object.

10.5. Aggressive and destructive tendencies (n+)

Aggressive and destructive tendencies are not often expressed in MIM sentence completions. Nevertheless, a special category is provided for these responses. Such tendencies have similarities with negatively valued motivational goals that are positively wanted ($n+$). We, therefore, code aggressive and destructive tendencies in terms of $n+$. But in order to distinguish them from positive tendencies towards negatively evaluated goals in general, we now put the symbol $n+$ in front of the code. Thus, the symbol $n+$ at the beginning of a code is the distinctive sign of aggressive or destructive tendencies. For instance, the goal: *I want to destroy my good reputation* refers to a personality aspect of an egocentric social-contact type (the personal reputation in the eyes of others) and destroying it is a kind of social self-destruction = $n+S(Ceg)$. The goal: *I want to offend my friend* (C,a) refers to a social contact with a friend. In both examples the destructive nature of the tendency is coded in terms of $n+$ preceding the content symbol; the latter is coded $n+C,a$. Self-aggression is coded $n+S$; aggression directed towards others is coded $n+C$. The destruction of objects (possessions) belonging to others is coded $n+P(C,...)$. Physical self-destruction is coded $n+Spre(ph)$. Destructive opposition against all restraints or restrictions of personal autonomy is coded by adding the symbol for barriers (b) to $n+$. Thus, $n+(b)$ preceding $Saut$ refers to destructive opposition to restraints (barriers) for personal autonomy: $n+(b)Saut$.

Aggression that is manifested in play behavior is coded $n+L$.

11. Responses to negative inducers (-)

Responses to negative inducers are coded following the general rules of the content code. The object that is feared is coded with the same symbols as for positive goal objects. The minus sign preceding the code indicates that the goal object was expressed to a negative inducer.

Example: *I don't want to fail in my exams = -R 3*.

When the goal object is NOT to do, to be, or to have something, the lower-case n precedes the symbol for the main category, in the same way as for responses to positive inducers (cf. supra, p. 162).

Example: *I wouldn't like...to work in a factory = -nR 2*.

The desired object in this example is 'not to work in a factory' (nR_2). We explained already that the minus sign does not refer to a double negation, it does not affect the content of the code.

12. Motivations related to the test (Tt)

Some subjects express motivational goals that are related to the ongoing research itself or to their responding to the 'test'. This kind of response may be interesting for the researcher, informing him about the subject's attitude towards the test. Moreover, these responses are important with regard to Time Perspective Analysis (cf. p. p. 108).

We distinguish three types of test-related responses:

Tt	responses expressing a neutral attitude or consisting of irrelevant or unclear remarks;
$Tt+$	responses with a positive attitude towards the test or the research;
$Tt-$	responses with a negative attitude towards the test or the research.

13. Unclassifiable and blank responses (U)

Besides inducers to which a subject does not respond, there are several types of unclassifiable responses. If this category of items attains a high proportion, it may be an interesting symptom for the category of subjects under investigation. With normal groups, subjects giving more than 10 percent of such responses will be eliminated. Some of these responses may reveal the attitude of the subject or some other characteristics such as the subject's difficulties in understanding some inducers, irrelevant associations produced by the inducers, etc. They may also contain useful information for the understanding of other responses. With this purpose in mind, this category of response can be divided in the following subcategories:

U unclassifiable without specification;

$U?$ understanding of the response is uncertain;

Ub blank (no response given);

Ui irrelevant with regard to the inducer or no motivational content (statement of fact, etc.).

Obviously these U-responses are not to be used in the motivational classification system itself. In the statistical tables referring to the subjects' motivations in each category, it will be sufficient to mention the global number of items in Tt and U, or even the total of Tt and U together. A qualitative analysis of these responses can sometimes lead to interesting conclusions.

Conclusion

The code for motivational content analysis is constructed in such a way that the different components or elements of each expressed motivational goal can be distinguished and analyzed. For example, the desire *to know and understand one's own abilities better* has two components: first a cognitive motivation is involved, and second a motivational preoccupation with oneself. The first element is coded in terms of E, the second in terms of S. The com-combination of these symbols gives the code for that motivational object = $E(S)$. The codes allow an analysis of the different elements implied in the object. The same holds for most other motivational categories (social, professional, religious, cognitive, recreational, etc.). A simple addition of each category of symbols or the frequency with which each component is represented in the global codes can be used as an index of the relative importance of each motivational component. The researcher decides in accordance with the goal of his research how to combine symbols in a meaningful motivational component. Thus, for instance, the symbol Ceg implies an S-element since it refers to egocentricity.

The code for people and the code for modalities can both be used for a more detailed analysis of the subjects' network of social relations and for motivational aspects. Due to its flexibility, the coding technique can easily be adapted for the specific requirements of the research. The capacity of the content codes as well as the time perspective codes to be processed by computer greatly facilitates statistical analysis (see *Appendix C*).

IV. LIST OF EXAMPLES

S
    to be happy;
a well balanced personality;
to be happy with my husband: $S(Cint,e)$;
to have an unhappy life: $-S$;
never to be unhappy again.

Sc
    to have a certain personal value;
to be someone in my own eyes;
that I very often feel inferior: $-Sc$.

$Spre(ps)$
    to regain my calm;
that something unexpected would destroy my life: $-Spre(ps)$;
not to be destroyed by the criticism of others: $Spre(ps)(np)$;
not to be anxious any more: $Spre(ps)(np)$;
to become disappointed in my expectations: $-Spre(ps)$.

$Spre(ec)$
    to be sure of my daily income;
to have a stable economic situation;
not to lose money in an investment: $Spre(ec)(np)$.

$Spre(ph)$
    to stay healthy;
to become healthy again;
to have an accident: $-Spre(ph)$;
to get rid of my cold;
not to be sick any more: $Spre(ph)(np)$;
not to drown: $Spre(ph)(np)$.

$Spre(T)$
    that I will die one day: $-Spre(T)$;
to survive in my children;
not to disappear completely after my death: $Spre(T)(np)$;
that I cannot live forever: $-Spre(T)$.

$Saut(ps)$
    to be free;
to do what I want to do;
to have my own personal ideas;

168 Manual: Motivational Content Analysis

	not to have to render an account: *Saut(ps)(np)*;
	not to be influenced by the opinions of my husband: *Saut(ps)(np)*;
	not to care about the opinion of others: *Saut(ps)(np)*.
Saut(ec)	to earn my own living;
	to be independent of my parents;
	that I always must ask my parents for money: *-Saut(ec)*;
	to have a job so that I'll be financially independent of my husband: $R_2 \rightarrow$ *Saut(ec)*.
Saut(ph)	to keep house myself (older person);
	to take care of myself;
	that I cannot walk by myself: *-Saut(ph)*;
	that my wife must help me to get out of my bed: *-Saut(ph)*.
Scar	to be a realist, an idealist, a volunteer;
	to control myself more;
	to be devoted, humble, indulgent, courageous;
	that I am sometimes stubborn: *-Scar*.
S(Calt)	to be open to others, to be friendly, to be social, to be helpful;
	to be honest.
S(Ceg)	to be popular; to have power or influence;
	to be an important person; to be famous;
	to have a great reputation;
	to be well known; to be someone special.
Sph	to be beautiful;
	to be slim;
	to lose some weight;
	that my hair are already grey: *-Sph*;
	to be a well-dressed lady;
	to take care of my hair-dress;
	that I have a scar in my face: *-Sph*.
Sapt	to be intelligent, skillful, clever, etc.;
	to be able to speak French: *Sapt(Ew)*

that I cannot sing: $-Sapt(L)$;
that I could type faster: $Sapt(R2)$;
that I cannot swim: $-Sapt(L,ph)$.

SR	to realize my dreams, my goals, my plans, etc.; to succeed in life; to lead a happy and calm life; to have a beautiful life.
SR2	to become a plumber, a lawyer, a doctor, etc.; to become a scientist: $SR2(E)$; to become successful in my career; not stay a secretary for the rest of my life: $SR2(np)$.
SR3	to become a real university student; to stay a strongly motivated student: $SR3car$.
SRcar	to stay humble in my life; to live a decent life.
SR(Calt)	to live for the benefit of others; to devote my life to my family; to dedicate my life to the poor.
SR2(Calt)	to dedicate my professional life to the education of mentally handicapped children.
SR(Ceg)	to become a leader; to become a star; to become someone who is indispensable for my country; to become an important person in social life.
SR(ph)	to become Miss America, a beauty queen; to stay a good-looking woman.
SRaut(ps)	to organize my own life; to live as I want to live; a free and independent life.
SRaut(ec)	to become financially independent; to become selfsupporting.

170 Manual: Motivational Content Analysis

$SR(P)$	to become a rich man.
$SR(Tr)$	to dedicate my life to God; to become a priest.
R	to do something important; to work in my garden; to wash my car; to fix the roof of my house; to realize a plan or project.
R_2	to find a job; to teach; to help my patients; to lose my job: $-R_2$; to stay without a job: $-R_2$; not to stay jobless: $R_2(np)$.
R_3	to go to classes; to study; to write my thesis; to succeed in the exams; to continue my studies; to apply at the university of...; not to fail in the exams: $R_3(np)$; to fail: $-R_3$; to finish college: R_3.
R_3a	to succeed with honors; to obtain a higher GPA than last year.
$R_3(m)$	to study hard, always, very much.
nR	not to work in my garden this weekend; not to do anything tonight.
nR_2	not to work for that company; that I must help my father in his work instead of going to college: $-nR_2+R_3$ or $R_3(b)$ (cf. p.161).
nR_3	that I must study philosophy: $-nR_3$;

	not to go school; that I must always study: *-nR3*; not take summer courses.
nR(pb)	to sleep long; to sleep a whole day long.
nR(ps)	to find peace; to be calm finally.
C	to write a letter to my friend: *C,a*; to have many friends: *C,a*; to go out with a girl: *C,e*; to be alone: *-C*; not to meet Albert: *nC,e*; not to meet anyone: *nC,g*.
Cint	to marry: *Cint,e*; to find my true love: *Cint,e*; to see my parents again: *Cint,p*; to have a good friend to whom I can entrust something: *Cint,a*; to stay with my wife: *Cint,e*.
Calt,.../R	to help my friend: *Calt,a/R*; to help my father in the garden: *Calt,p/R*; to give money to the poor: *Calt,b/R*; to give my children a good education: *Calt,p/R*; to work for my people: *Calt,pl/R*; to take more care of my children: *Calt,p/R(d)*.
C(an)	to take care of my dog.
Ceg,.../R	to convince my friend that I am right: *Ceg,a/R*; to take control of our group: *Ceg,gr/R*; to show others who I really am: *Ceg,g/R*.
C(Tr)	to meet God; to go to heaven and see God.
C(Tx)	to be in contact with the ultimate.

172 Manual: Motivational Content Analysis

$C_2(af)$ — the love and friendship of my wife: $C_2,e(af)$;
that my children will love me: $C_2,p(af)$;
that my victim would forgive me: $C_2(af)$.

$C_2(ap)$ — that my qualities be recognized: $C_2(ap)$;
that my parents would appreciate my work: $C_2,p(ap)$;
that my colleagues underestimate me: $-C_2,gr(ap)$;
when my boss asks for my opinion: $C_2,si(ap)$.

$C_2(su)$ — that my father does not help me: $-C_2,p(su)$;
that my husband does not help me in the kitchen: $-C_2,e(su)$;
that my teachers would help me more: $C_2,si(su,R_3)(d)$;
that my father refuses to pay my vacation: $-C_2,p(su,L)$;
a little more encouragement from my wife: $C_2,e(su)(d)$.

C_2 — that my father would give me a car: $C_2,p(P)$;
that my parents would allow me to go to Europe for vacation: $C_2,p(L,Ew)$;
that my brother would give me a stethoscope: $C_2,f(P,R_2)$.
that they are informed about my things: $-nC_2(E)$;
that they would leave me alone: $nC_2(R)$.

C_3 — that my brother would be more intelligent: $C_3,f(Sapt)$;
that my father would recover soon: $C_3,p(Spre,pb)$;
that my friends would be good persons: $C_3,a(Scar)$;
when my sister will be more independent: $C_3,f(Saut,ps)$;
that my children would suffer injustice at school because I am a criminal: $-C_3,p(ap)$.

C_3,H — more justice in the world;
that segregation would not exist anymore;
freedom in the world;
the development of third world countries;
that the revolution would make progress;
social inequality: $-C_3,H$.

E — to know everything;
to understand everything.

$E(S)$	to know myself; to know what I want to be (to become); to find my true self.
$E(R_2)$	to know more about my job; to know how that new machine works.
$E(C)$	to understand my parents: $E(C,p)$; to get to know our neighbors better: $E(C,gr)$.
$E(l)$	to have more life experience; to experience new things.
$E(w)$	to know the habits and customs of...; to know more about classic music: $E(w)(d)$; to see the Grand Canyon; Niagara Falls; to visit the Louvre.
$E(Tr)$	to know more about God: $E(Tr)(d)$; to find an answer to my religious questions: $E(Tr)$.
$E(Tx)$	to find the meaning of life; to find the ultimate truth.
Tr	to find eternal happiness; to pray; to go to mass.
Tx	that life is meaningless: $-Tx$; not to exist: $Tx(n+)$; that life is worth the trouble: Tx.
P	to buy a lot of land; to have more money: $P(d)$; to have more comfort in my kitchen: $P(d)$.
$P(Spb)$	to buy clothes, make-up, a wig, etc.
$P(R_2)$	to buy a new stethoscope; to buy a new typewriter.

174 Manual: Motivational Content Analysis

P(R3)	a study grant; a small library in my field of study.
P(L)	a sportscar, a tennis racket, a yacht, etc.
P(Ew)	to buy a globe; to have some classical records; to buy an encyclopedia.
P(an)	to have a dog, an aquarium.
L	to make a nice tour; to take vacation; to relax; to go out; to watch a football game, etc.
L(E)	to listen to music during the weekend; to read an historical novel; to play the piano; to visit the Louvre during my vacation in Europe.
L(pb)	to go swimming; to go for a walk; to play tennis; to relax by gardening.
L(nR)	to relax by doing nothing; to enjoy a day of dolce farniente.
L(C,e)	to go dancing.
L(ss)	that the weather would be nice; to have sex; to smoke a cigarette; to drink a cup of coffee; that it rains so much: -L(ss).
L(gf)	to gamble in a casino; to play lotto.
Tt+	to help you with this test; to answer this test honestly.

Tt-   that I agreed to take this test: *-Tt-*;
to finish this test as soon as possible.

APPENDIX A

MIM INSTRUCTIONS AND LIST OF INDUCERS

I. SUGGESTIONS TO THE INSTRUCTOR

The three lists of MIM inducers or sentence beginnings are printed below. The format of the booklets we use is shown in Fig. 4. Each inducer is typed on the upper left hand corner of a numbered page.

Fig. 4. Format of the MIM booklets as currently used
(10 cm x 5 cm)

The 40 pages with the positive inducers are stapled together and form a little booklet; the 20 negative sentence beginnings form a second booklet. On the cover page of the first booklet, the subjects are requested to give their age, sex, profession, or school grade. After having read the instructions, they write the first sentence completion on page one and immediately go to the following page. It is important that each inducer and each sentence completion be written on a separate page in order to prevent the subject from seeing the whole list of inducers and his previous responses. After finishing the test,

178 Appendix A

the subject puts his two booklets in one envelop in order to keep them together. The MIM is anonymous. The information requested on the cover page should not threaten this anonymity.

Two shorter versions of the MIM contain only 30 and 15 or 20 and 10 inducers, respectively (see below).

Although the MIM is limited to the investigation of conscious motives, it invites subjects to express also motivational objects that belong to the intimate personality level, whether or not they are socially acceptable. For this purpose, the instructions insist on anonymity and frankness.

In the case of group sessions, each subject should be seated a comfortable distance from the other subjects, and school teachers or others in position of authority should not be present.

Individual instruction sheets are distributed together with the booklets and read aloud by the instructor. The text of these instructions is given below. If necessary, it is to be adapted to the educational level of the subjects and to the special requirements of the research.

It has been found that deeper motivational levels can be investigated when the instructions invite the subjects to consider seriously what they really want or are striving for in their life, instead of just asking them to express motivations that come to their mind. Such an inquiry could be done - if deemed desirable - with a short list of inducers after the first administration in the usual way.

As stated above (see p. 63), the MIM inducers can also be used in a modified form to investigate several aspects of a person's motivation. For example: *As a housewife, I intend...; As a Jew, I want...; As a Huron Indian, I would oppose it if...,etc.*

French, German, and Dutch versions of the MIM inducers and instructions are given in the French edition of this book (Nuttin, 1980).

II. INSTRUCTIONS TO THE SUBJECTS

Please read this carefully

On each page of the two booklets that you have before you, you will find some phrases, for example, "I wish...", "I fear...". These phrases constitute the beginning of a sentence. You are asked to complete this sentence on the same page of the booklet by applying the words to yourself. Thus, in the first example given, you continue the sentence by indicating something which you *personally*, really wish.

You need not think for a long time about each sentence; simply write whatever comes to your mind when you apply the words on each page to yourself.

The essential thing is that you mean what you write. It is not a question of constructing a grammatical sentence, but of *expressing the real objects of your wishes, plans, intentions*, etc., on each page. Try to reply in such a way that each sentence has a meaning by itself.

If a phrase ends in one or two words within parentheses, for example, "*I wish (that or to)...*", then you may finish it by using either of these words. In the example, therefore, you may say "*I wish that...*" or "*I wish to...*", whichever statement seems to express your wish more accurately.

We understand that people don't like to tell others what they wish or fear. This is why you are asked to reply *without giving your name*. It is not important for our research that we know the name of the person who completes the sentences, but it is essential that he does it sincerely and personally. It is absolutely guaranteed that no one will try to identify the person who made these responses. The experimenter has arranged to insure that your booklet will not be identified. Therefore, you should not hesitate to express your most intimate wishes or fears, even when they may seem socially unacceptable. The only condition is that they exist in you or come to your mind from time to time.

You may have the impression that similar phrases are repeated on several pages. This is to give you the chance to express many objects of motivation. When you have such an impression, you should not try to remember what you have written before, but simply write what comes to your mind when you read the words on the new page.

Do not change the words printed on each page.

One last remark. You have been asked to take part in a psychological study. This research is valid only if the questionnaire is completed truthfully and intimately. If you are not willing to do this seriously, please return the booklets blank.
Thank you for your cooperation.

III. COMPLETE LIST OF MIM INDUCERS

1. Positive inducers

1. I hope...
2. I am working towards...
3. I intensely desire...
4. My greatest satisfaction is (*or* will be)...
5. I intend to...
6. I wish...
7. I try to...
8. I long (for, *or* to)...
9. I am determined to...
10. I will be glad when...
11. I wish very much to be able to...
12. I am dreaming of (not at night)...
13. I definitely have the intention to...
14. I want...
15. I would not hesitate to...
16. I am striving (to, *or* for).
17. I have a great longing (for, *or* to)...
18. My greatest reward is (*or* will be)...
19. I strive to...
20. I would like to be able to...
21. I would like so much...
22. I am trying (to, *or* for)...
23. I am preparing myself (to, *or* for)...
24. I am resolved to...
25. I will be very happy when...
26. I ardently desire...
27. I am ready (to, *or* for)...
28. I will do everything possible to...
29. I would like very much to be allowed to...
30. I am doing my best to...
31. I will be most satisfied when...
32. I hope with all my heart to...
33. I would spare nothing (to, *or* for)...
34. I yearn for...
35. I am inclined to...
36. At all costs, I am willing to...
37. My intense hope is...
38. Quite definitely, I want...
39. As soon as possible, I would like...
40. Quite strongly I strive (to, *or* for)...

2. Negative inducers

n1. It would displease me very much if...
n11. I really don't want...
n12. I am worried that...

n2. I don't want...
n3. I would oppose it if...
n4. I think it is sad that...
n5. I would not like it if...
n6. I try to avoid...
n7. It displeases me that...
n8. I would not in any way like it if...
n9. I am afraid that...
n10. I would regret it very much if...
n13. I wouldn't like...
n14. I don't like to think that...
n15. I find it unbearable that..
n16. I am not inclined to...
n17. I am sorry (that)...
n18. It would annoy me very much if...
n19. I fear that...
n20. I would not want...

IV. SHORTER FORM A OF MIM INDUCERS(1)

1. Positive inducers

1. I hope...
3. I intensely desire...
5. I intend to ...
6. I wish...
7. I try to...
8. I long (for, *or* to)...
9. I am determined to...
10. I will be glad when...
12. I am dreaming of (not at night)...
13. I definitely have the intention to...
14. I want ...
15. I would not hesitate to...
16. I am striving (to, *or* for)...
17. I have a great longing (for, *or* to)...
19. I strive to.....
20. I would like to be able to...
21. I would like so much...
22. I am trying (to, *or* for)...
23. I am preparing myself (to, *or* for)...
24. I am resolved to...
25. I will be very happy when...
26. I ardently desire...
27. I am ready (to, *or* for)...
28. I will do everything possible to...
29. I would like very much to be allowed to...
30. I am doing my best to...
32. I hope with all my heart to...
33. I would spare nothing (to, *or* for)...
35. I am inclined to...
40. Quite strongly I strive (to, *or* for)...

2. Negative inducers

n1.	It would displease me very much if...	n11.	I really don't want...
n2.	I don't want...	n13.	I wouldn't like...
n3.	I would oppose it if...	n16.	I am not inclined to...
n4.	I think it is sad that...	n17.	I am sorry (that)...
n5.	I would not like it if...	n18.	It would annoy me very much if...
n6.	I try to avoid...	n19.	I fear that...
n9.	I am afraid that...	n20.	I would not want...
n10.	I would regret it very much if...		

(1) The Form A has 30 positive and 15 negative inducers. Each inducer keeps the number it has in the complete list.

V. SHORTER FORM B OF MIM INDUCERS(1)

1. Positive inducers

1. I hope...
3. I intensely desire...
5. I intend to...
6. I wish...
8. I long (for, *or* to)...
10. I will be glad when...
14. I want...
16. I am striving (to, *or* for)...
17. I have a great longing (for, *or* to)...
20. I would like to be able to...
21. I would like so much...
22. I am trying (to, *or* for)...
24. I am resolved to...
25. I will be very happy when...
26. I ardently desire...
28. I will do everything possible to...
29. I would like very much to be allowed to...
30. I am doing my best to...
32. I hope with all my heart to...
40. Quite strongly I strive (to, *or* for)...

2. Negative inducers

n1. It would displease me very much if...
n2. I don't want...
n3. I would oppose it if...
n5. I would not like it if...
n6. I try to avoid...
n9. I am afraid that...
n10. I would regret it very much if...
n11. I really don't want...
n13. I wouldn't like...
n20. I would not want...

(1) The Form B has 20 positive and 10 negative inducers. Each Inducer keeps the number it has in the complete list.

APPENDIX B

INVENTORY OF MOTIVATIONAL CATEGORIES
and
INVENTORY OF MOTIVATIONAL OBJECTS

In addition to and based on the Motivational Induction Method (MIM), two other techniques were developed to get samples of motivational objects and to study their subjective intensity. Both are of the inventory type. While the MIM avoids suggesting anything to the subjects, the two inventories provide them with a large variety of motivational categories and objects that orient their attention to different motivational areas.

In spite of the rather high number of inducers in the MIM and the instructions asking the subjects to consider the whole range of their activities and goals, some subjects tend to limit their sentence completions to a restricted number of domains (such as professional life, studies, and family life). This tendency may be considered as a certain kind of perseveration, but it does not prevent us from getting representative samples of motivational objects as long as *groups* of subjects are studied. In fact, it may be expected that the perseveration of one subject in a specific field will be compensated by an emphasis on other fields by other members of the group. In studying *individuals*, however, this tendency may result in less reliable data. Therefore, the Inventory of Motivational Categories (INCAM) can be recommended for the study of motivational content and future time perspective in individual subjects or in very small groups. It can be administered after the MIM and could also replace it, if necessary. The motivational objects thus obtained can be analyzed in terms of the code for future time perspective and/or the code for motivational content analysis, as explained above. The Inventory of Motivational Objects (INOM) was developed to measure the subjective intensity or strength of different motivational concerns, as will be shown below.

I. THE INVENTORY OF MOTIVATIONAL CATEGORIES (INCAM)

The Inventory of Motivational Categories (INCAM) consists of a sheet of paper with a list of motivational categories and subcategories referring to different objects and possible goals. Since the list has been derived from the content analysis of our MIM data, it is not an a priori classification or enumeration of motivational objects. The categories and objects mentioned were actually found in the data of groups of subjects.

In administering the INCAM, we usually proceed as follows.

The list of motivational categories (see p. 187) and a sheet of paper with the following instructions are given to each subject. A few blank sheets are added.

Instructions (1)

In the preceding questionnaire you expressed various objects that you desire, hope or fear to obtain, that you are planning or striving for, etc. However, it is quite possible that many other objects that you also desire or fear did not spontaneously come to mind when completing the sentences.

Therefore, we now give you a second questionnaire. In it are mentioned different kinds of objects related to the motivational concerns of many people; e.g. "personal characteristics" "possessions", etc. Indeed, many people want or fear a certain characteristic, want to own or to have certain objects, etc.

We ask you to read through this list very carefully, point by point, and to ask yourself whether you have some personal desires or fears in the field mentioned or in a related field for each of these points. Write down these concrete wishes or fears on the extra sheets of paper that are joined to these instructions. Start a new line for each motivation. Whether or not you have mentioned these wishes or fears in the preceding questionnaire is of no importance at all. Write down what comes into your mind as a desire or as a fear when you consider the motivational field indicated by each point in the list. The only important thing is that you do not invent something at the moment, but that you write down what you really want or fear.

Do not give vague or general responses, but concrete objects of wishes or fears. You don't have to formulate full sentences; it is sufficient to indicate the objects themselves as concretely as possible. When the object mentioned is something you wish, write the sign + in front of the word or sentence; when it is something you would like to avoid, write a sign - before the sentence.

The order of motivational objects as you write them down does not have to correspond to the order of the points or categories in the list. That is why we did not number the categories. When an important motivation, related to an earlier mentioned category, comes up in your mind, you may write down that motivation on the following line. The order is of no importance, but make sure that you read everything that is in the list very carefully. In case you have no motivations in a certain category, just go on to the next category.

TABLE 11

List of categories as presented to the subjects

What do you want, hope or fear concerning yourself as a person:
- *intellectual abilities and skills;*
- *moral or characterological qualities;*
- *social qualities and relationships with others, social standing, prestige;*
- *freedom and autonomy;*
- *subsistance;*
- *physical and external characteristics;*
- *yourself in general, equilibrium, adaptation, etc.;*
- *something else for yourself?*

Concerning your accomplishments, realizations or activities:
- *what would you like to do or to become;*
- *all wishes or fears related to your job or your activities, now or later;*

Concerning your social contacts or social life:
- *social contacts in general, friends, being alone, hostility, etc.;*
- *relationship with parents, family, home, etc.;*
- *more intimate relations: love, eroticism, sexuality, marriage, founding a family, your own children, etc.;*

Do you wish something for some other people, groups, communities, for a certain cause or ideal? - what?

Do you wish to receive material or moral goods (such as consideration, love, help, etc.) from others?

Do you want some material possessions for yourself:
- *one or other object for now or later, in any field (sports, profession, arts, culture, physical appearance, etc.);*
- *prosperity in general, comfort, etc.;*
- *to have a pet;*

Do you have wishes or motivations related to your diversion or other such activities? Which one? (competitive games, social or cultural diversion, traveling, hiking, sensual or sexual pleasure, vacation, etc.);

Do you have wishes related to your instruction, information, and knowledge (with regard to people, nature, cultures, life experiences etc.)?

Do you have wishes or motivations related to religion?
- *religious or transcendental values, philosophy of man, philosophy of life;*
- *your religious life, piety, virtue;*
- *apostolic activities, your good works;*

Probably there are other things that you wish, are planning for, or fear and that are not on this list. Please concentrate once more and write down on your sheets of paper whatever comes into your mind.

In order to test the hypothesis that subjects will express a greater variety of motivations in the Inventory than in the MIM, the two techniques were administered, successively, to the same group. Ninety subjects (45 male and 45 female undergraduate students) first responded to the MIM and, immediately afterwards, to the Inventory. The total number of motivational objects expressed in both techniques and the various categories to which these motivations belong were calculated. As hypothesized, the average number of motivational categories covered is higher in the Inventory: while in the MIM the motivations of most subjects are spread over from five to seven categories out of eight (with an average of 6.5), practically all categories are covered in the Inventory, namely seven or eight categories with an average of 7.5. Considering only those categories in which at least 50 percent of the subjects express some motivations, it is found with the Inventory that this is the case for each of the eight categories, whereas only six out of eight categories meet that criterion with the MIM (here the categories "Transcendental" and "Possessions" T and P do not meet the criterion).

As to the total number of motivational objects expressed by our subjects, an opposite tendency was found: 90 percent of the subjects expressed a higher total number of motivational objects with the MIM than with the Inventory. The average is 63.65 and 51.59, respectively. This shows that with the MIM, on the average, more than one motivational object was expressed for each inducer (an average of 63.65 objects for 60 inducers). In the Inventory, however, there is no pressure from inducers, and an average of only 51.59 motivational objects was expressed. This lower number of motivations is spread over a larger number of motivational categories, as said before.

The relative frequency of motivational objects in the different main categories has also been calculated. As shown in Table 12, it was found that in both the INCAM and the MIM the categories C and S attract the highest proportion of concrete motivations (respectively 32 percent and 24 percent with the MIM, and 33 percent and 32 percent with the Inventory). Motivations belonging to the main category R (including objects that belong to the subjects' study activities) are frequently expressed with the MIM (20,9 percent), but much less frequently with the INCAM (4,8 percent). The obvious explanation of this finding is that this kind of motivations (related to study) comes up frequently in the mind of our subjects when presented with a series of inducers, whereas it is only mentioned once in the Inventory. However, this difference in the frequency of expression appears to have its psychological meaning. Indeed, it was found in another study that the subjective intensity if this frequently expressed type of R_3 motivations is very strong (see Table 13). The opposite holds for objects in categories P (possessions) and L (leisure). The small proportion of these categories of motivations expressed in the MIM corresponds to their low degree of intensity as shown also in Table 13. This finding tends to confirm what has been said above (cf. p. 64), that, in the MIM, a high frequency of expression of a certain type of motivation is an indication of its intensity. The high frequency of P and L motivations expressed in the Inventory is related to the fact that some subjects tend to enumerate quite a number of concrete objects they want to posses, or things they want to do as recreation. Both categories, however, have such a low degree of intensity (at least in these subjects) that these motivations do not spontaneously occur to them as responses to the motivational inducers.

TABLE 12

Relative frequency of responses in the main categories of the MIM and the INCAM

MIM		INCAM	
Categories	Percent	Categories	Percent
C	31.9	C	33.5
S	24.3	S	32.0
R	20.9	L	7.6
SR	9.9	P	7.2
L	3.5	E	6.8
E	2.1	SR	5.3
T	1.3	R	4.8
P	0.5	T	1.9
Other	5.6	Other	0.9

II. THE INVENTORY OF MOTIVATIONAL OBJECTS (INOM) AND INTENSITY MEASURES

The second Inventory is also derived from MIM data. It consists of a list of concrete motivational objects as expressed by university students and taken from the different main categories and subcategories of motivations in our content analysis. It was constructed for the purpose of measuring the subjective intensity of different types of motives. The motivational objects as expressed by our subjects have been formulated in somewhat more general terms in such a way that they are applicable to several specific forms of motivation within the subcategory. For instance, "to be myself" takes many different forms in several subjects, but everybody can recognize in it his own form of being himself.

A large variety of fifty categories and subcategories of motivations are represented in the Inventory. It is divided into two parts: 105 objects of positive motivation and 55 objects of negative motivation (see list below). The more important categories of our content analysis such as *Self* and *Contact* are represented by a higher number of subcategories than is the case for, say, categories *T* or *P*. For each of these subcategories, two positive and one negative motivational object were chosen. The number of items within each main category corresponds to the relative frequency of motives spontaneously expressed in that category by our subjects. A few items from Murray's list (1938) have been added to the motivational objects taken from the MIM data, because some subcategories were underrepresented in our sentence completions. Moreover, five positive and five negative motivational objects are mentioned twice in our lists in order to check the reliability of the subjects' intensity measures. The twice-mentioned items are numbers 7-101, 11-102, 20-104, 74-103, and 84-105 among the positive motivations, and numbers n6-n55, n8-n52, n18-n43, n34-n51, and n40-n54 among the negative ones.

The subjects are given the lists of items (see p. 194) and the following instructions:

Instructions

On the lists you just received, there are a number of short phrases indicating goal objects that are pursued with more or less intensity by some people, while others are not interested in some of these goals. Thus, for instance, the first item says "...to be myself"; which means: *I would like*, or *I want* "to be myself". The third item says..."to be physically attractive", which means: *I want*, or *I would like*...etc. We invite you to read each of these phrases carefully and to ask yourself, for each of them, whether this is something *you personally* want or not; in other words, is it something which you are motivated for. This means that you are willing to make an effort to get it or it leaves you completely indifferent. In the latter case, you may say that that motivation is *absent* in your life and you put the number 0 in the left column beside that item. When, however, you are very or extremely motivated to obtain the goal expressed in a phrase, you put the number 6 in the same column beside that item. Thus, you have at your disposal 7 numbers to indicate the *degree of intensity* of your

192 Appendix B

motivation for each goal object. The meaning of the numbers is as follows:

0 = absent (= that motivation is *absent* in my life)
1 = very weak (= that motivation exists, but only in a very low degree)
2 = weak
3 = moderate
4 = rather strong
5 = strong
6 = very strong (= that motivation is something I am striving for with greatest intensity (2).

In the second list, you will find goal objects which are usually NOT wanted but avoided by people (Negative items). For instance: "...to fail in my studies". Read each of these phrases as if they were preceded by the words "I do NOT want to...", "I would NOT like that..."; "I dislike that...". Please put one of the numbers 0 to 6 beside each of these negative goal objects, according to the degree of intensity with which you dislike or avoid these objects. Remember that putting number 0 for the phrase "...to fail in my studies" would mean that a dislike for failing in your studies is absent, i.e. does not exist in your personal case; if the aversion for failing is extremely strong, you write number 6. Be careful to respond with the number corresponding to *your personal feeling.*

The list of positive and negative motivational objects is given in Table 14. The subject is presented with the items in that order. On the left, a column is provided for the subjects' response. For the reader's information, the following data are added: the letter-symbol following each item indicates the category or subcategory of our content analysis code to which the item belongs. The following number is the subjective intensity index (on a scale from 0 to 6) as obtained for each item by a group of 30 male and 39 female university students. Note that in our group of subjects the most intense positive motive is *to be myself* (index 5.1.), whereas *to be rich* and *to have no contacts with certain people* are the weakest ones. The average intensity of positive and negative items for each of the main categories of motivations is given in Table 13.

It may be added that some significant differences appear between male and female subjects. Motivational objects in the field of professional activity (including studies) are pursued with more intensity by females than by males ($p < 0.05$ for positive objects, and $p < 0.02$ for negative ones). Motives in the field of social contact (Category C) are also more intense in females.

TABLE 13

Subjective intensity of positive and negative motivational objects of different categories. Intensity scale from 0 (absent) to 6 (very strong).

Categories*	Posit.Mot.		Negat. Mot.	
R 3	1.	4.19	2a	4.28
C 3	2.	4.01	2b	4.28
S	3.	3.98	5.	3.79
SR	4.	3.93	1.	4.37
R	5.	3.54	4.	4.03
C	6.	3.35	7.	3.29
C 2	7.	3.21	8.	2.98
L	8.	2.97	9.	2.82
E	9.	2.81	6.	3.48
T	10.	2.57	10a	2.58
P	11.	1.62	10b	2.58

* For the meaning of the categories, see supra p. 141.

TABLE 14

List of positive and negative motivational objects of the Inventory of Motivational Objects (INOM)

A. Positive items

1. ...to be myself (S; 5.1)
2. ...to do my best (S; 4.2)
3. ...to be physically attractive (S; 2.8)
4. ...to succeed in life (SR; 4.6)
5. ...to obtain a powerful position (SR; 1.4)
6. ...to become an autonomous person (SR; 4.0)
7. ...to succeed in my exams (R_3; 4.8)
8. ...to have an interesting job later (R_2; 4.8)
9. ...to obtain what I have in mind (R; 4.9)
10. ...to make many social contacts (C; 4.4)
11. ...to be in love (C; 3.3)
12. ...to be respected (C_2; 3.5)
13. ...a better world (C_3; 4.8)
14. ...that a certain person would perform more (C_3; 2.6)
15. ...social improvements for everybody (C_3; 4.6)
16. ...to be informed about everything (E; 3.5)
17. ...to achieve something in arts (E; 2.8)
18. ...to inherit eternal life (T; 1.9)
19. ...to have beautiful clothes (P; 1.6)
20. ...to relax (L; 3.5)
21. ...to enjoy myself in good company (L; 3.1)
22. ...to stay healthy (S; 4.7)
23. ...to be intelligent (S; 4.0)
24. ...to prepare myself for a harmonious family life (SR; 4.1)
25. ...to build up a good position later (SR; 2.9)

26. ...to make good progress in my studies (work) ($R3$; 3.8)
27. ...to meet many people (C; 4.5.)
28. ...to help others (C; 4.8)
29. ...to be understood ($C2$; 4.1)
30. ...that the others would succeed in their exams ($C3$; 3.7)
31. ...that the members of my family keep contact with each other ($C3$; 3.6)
32. ...to be a little further on in time (E; 1.8)
33. ...to cooperate in realizing truth and justice (T; 4.0)
34. ...to buy records and books (P; 2.5)
35. ...to have a good meal (L; 1.7)
36. ...to be master of myself (S; 3.9)
37. ...to defend myself against possible failures (S; 3.9)
38. ...to have more freedom (S; 3.2)
39. ...to realize my plans and projects (SR; 4.5)
40. ...a life in the service of other people (SR; 4.7)
41. ...to develop my capacities (SR; 4.4)
42. ...to obtain good results in the exams ($R3$; 4.5.)
43. ...to achieve something (R; 4.4)
44. ...to be together with the one I love (C; 5.0)
45. ...to assert myself (C; 2.7)
46. ...to give a certain person hell (C; 1.3)
47. ...that other people are happy ($C3$; 4.8)
48. ...that there is social equality for everybody ($C3$; 4.9)
49. ...that the academic year was over already (E; 2.8)
50. ...to be rich (P; 1.1)
51. ...to experience strong sexual satisfaction (L; 2.4)
52. ...to devote myself to a certain cause (S; 4.5.)
53. ...to understand other people (S; 4.8)
54. ...to be physically beautiful (S; 2.6)
55. ...to be good in my profession (SR; 5.0)

56.	...to enjoy spending some time in idleness (R; 2.4)
57.	...to make someone happy (C; 5.0)
58.	...to have no contacts with certain people (C; 1.1)
59.	...that others may succeed (C_3; 3.2)
60.	...to gain much scientific knowledge (E; 3.1)
61.	...to take more care of my prayers (T; 1.7)
62.	...to take a vacation (L; 3.2)
63.	...to be happy (S; 5.0)
64.	...to become a person caring for the needs of others (SR; 4.8)
65.	...to finish my studies (R_3; 4.7)
66.	...to tackle also difficult problems (R; 3.9)
67.	...to convince others of my opinions (C; 2.7)
68.	...to receive affection (C_2; 4.3)
69.	...that a certain person may succeed in life (C_3; 4.4)
70.	...to go out with friends now and then (L; 3.5)
71.	...to be good (S; 4.8)
72.	...to be friendly to others (S; 4.7)
73.	...to obtain a high social status (SR; 1.6)
74.	...to study (or work) hard (R_3; 3.1)
75.	...to match with someone else (R; 1.6)
76.	...to be alone now and then (C; 3.2)
77.	...to receive help (C_2; 2.7)
78.	...that people need each other more (C_3; 3.6)
79.	...to believe in God (T; 2.7)
80.	...to be outdoors in nature (L; 4.5)
81.	...to be independent of everybody (S; 2.3)
82.	...to have a high I.Q. (S; 3.0)
83.	...to become a good father or mother (SR; 4.7)
84.	...to obtain my diploma (R_3; 4.4)
85.	...to have a good rest (R; 2.7)
86.	...to defend myself (C; 2.0)

The inventory of motivational objects 197

87.	...not to be disturbed by others (C_2; 1.7)
88.	...to fill in this test honestly ($Test+$; 3.8)
89.	...beautiful boys or girls around me (L; 1.7)
90.	...going to live my own life (SR; 3.8)
91.	...to become an honest person (SR; 4.8)
92.	...to find a nice job later (R_2; 3.6)
93.	...to stop this test immediately ($Test-$; 2.3)
94.	...to obtain material possessions (P; 1.3)
95.	...to be appreciated (C_2; 3.3)
96.	...that someone loves me (C_2; 4.6)
97.	...that everybody would fill in this test well ($Test+$; 3.1)
98.	...that the others would leave me alone (C_2; 1.5)
99.	...to study languages (E; 2.9)
100.	...that this will be over soon ($Test-$; 2.5)
101.	...to succeed in my exams (R_3; 4.7)
102.	...to be in love (C; 3.5)
103.	...to study (or work) hard (R_3; 3.2)
104.	...to relax (L; 3.1)
105.	...to obtain my diploma (R_3; 4.4)

B. Negative items

n 1.	...to lack self-control (S; 4.3)
n 2	...to hurt others (S; 4.6)
n 3	...to be dull (S; 4.0)
n 4.	...not to be respected (C_2; 3.2)
n 5.	...never to become a professional man (SR_2; 4.2)
n 6.	...to neglect my studies (R_3; 4.0)
n 7.	...to fail in my intention (R; 4.2)
n 8.	...to be alone (C; 3.7)
n 9.	...to disappoint others (C; 4.4)

198 Appendix B

n10. ...to be criticized (C_2; 2.3)
n11. ...that there will be war again (C_3; 5.0)
n12. ...that a good friend would fail (C_3; 4.6)
n13. ...to be uninformed in many fields (E; 4.0)
n14. ...to be indifferent in religious matters (T; 2.7)
n15. ...to lose my vacation (my holidays) (I; 3.8)
n16. ...that this test would be useless ($Test+$; 3.3)
n17. ...that I do not oblige myself to sacrifice more (S; 3.1)
n18. ...to be dependent on others (S; 3.0)
n19. ...to become a failure (SR; 4.4)
n20. ...to become a bad parent (SR; 4.8)
n21. ...to fail in my studies (R_3; 4.6)
n22. ...to avoid difficult tasks (R; 4.1)
n23. ...that a certain relationship would break up (C; 4.6)
n24. ...to be indulgent (S; 1.8)
n25. ...that others do not leave me alone (C_2; 2.4)
n26. ...violent quarrels in the family (C_3; 4.0)
n27. ...that some people have more than others (C_3; 3.8)
n28. ...to have not much knowledge of human character (E; 4.4)
n29. ...that I'll have to lack my means of relaxation (P; 3.1)
n30. ...to inhibit my sexual impulses (L; 2.8)
n31. ...to become unhappy (S; 4.7)
n32. ...to be ugly or unattractive (S; 3.1)
n33. ...to become an asocial person (SR; 4.9)
n34. ...that I'll have to stop my studies (R_3; 4.4)
n35. ...that I cannot rest enough (R; 3.2)
n36. ...that I cannot assert myself now (S; 3.3)
n37. ...that others are unfriendly to me (C_2; 3.9)
n38. ...that others fail in life (C_3; 4.1)
n39. ...that time would proceed even slower (E; 1.9)
n40. ...to lose my religion (T; 2.5)

n41.	...to do without my refreshments or sweets (L; 1.6)
n42.	...to be unable to maintain myself psychologically (S; 4.2)
n43.	...to be dependent on others (S; 3.2)
n44.	...to become a bourgeois (SR; 4.8)
n45.	...that I'll have to do unsuitable work later (R_2; 4.6)
n46.	...to meet a certain person (C; 1.4)
n47	...that they ridicule me (C_2; 3.3)
n48.	...that I'll be deprived of my social entertainment (L; 4.1)
n49.	...to waste my time here ($Test$-; 2.2)
n50.	...to be without comfort (P; 2.0)
n51.	...that I'll have to stop my studies (R_3; 4.3)
n52.	...to be alone (C; 3.9)
n53.	...to fail in my exams (R_3; 4.3)
n54.	...to lose my religion (T; 2.5)
n55.	...to neglect my studies (R_3; 4.0)

NOTES

(1) This Inventory can also be administered to subjects who have not previously been given the MIM. In that case, the first paragraphs of the instructions have to be adapted and somewhat expanded.

If the *Subjective Intensity Index* is to be calculated (see the second Inventory), the subjects can be invited to add a symbol to each object (e.g. a number from 0 to 6) indicating the intensity of that motivation in their personal life. A Table with the meaning of these numbers should be placed in front of the subjects.

(2) In some of our researches with this Inventory, the numbers used were 1 to 7 instead of 0 to 6 as indicated above. In fact, it appeared to be easier for the subjects to use a zero (0), rather than the digit 1, for expressing the meaning that a given motivation was absent. Therefore, the numbers 0-6 are preferred to 1-7.

APPENDIX C

COMPUTER ANALYSIS OF MIM DATA

by W. LENS and A. GAILLY

In this section, we introduce a technique of transcribing the Motivational Content Codes and the Codes for Time Perspective in numerical codes and putting those codes on punchcards so that the information can be read by computer.

Two problems are discussed: 1. the transcription of the content and temporal codes in numerical codes (see Tables 15 and 16); 2. the arrangement of the numerical codes on a series of punchcards.

The number of cards (a set) needed for each subject depends on several factors that will be discussed further. It will become obvious that our technique can easily be adapted (extended or reduced) according to the research goal.

Card nr 1

The first three columns of each card are used for the identification number of the subject and the next two columns for the rank number of the card in the set of cards for that individual.

The remaining columns of the first card are used for personality data about the subject (sex, age, socio-economic level, education, profession, marital status, etc.)

Card nr 2 and following cards

These cards are used to store the numerical transcription of the content codes and temporal codes. The numerical transcription for the temporal codes is given in Table 15, and for the content codes in Table 16. It is not intended to be exhaustive. As noted, it can be extended or reduced according to the requirements of a study.

Appendix C

The first three columns of all cards referring to the same subject contain the same number. Columns 4 and 5 give the ranknumber of each card in the set of cards for one subject. The transcription of codes starts with the sixth column. Two practical rules must guide this operation.

1. A certain number of columns is needed for each coded goal object. This number depends on the amount of information one wants to transcribe on the punchcard. Fifteen columns are needed for each goal object if the maximal amount of information is used.

The 15 columns are used as follows:

- columns 1 and 2 for the number of the inducer of the response;
- columns 3, 4 and 5 for the content code;
- columns 6 and 7 for the code for people;
- columns 8 and 9 for the code for modalities;
- columns 10, 11, and 12 for the time perspective code;
- columns 13, 14, and 15 for references to the past and for some temporal modalities.

The number of columns used for each motivational object is constant within each study. When the information to be punched in some of the columns is absent, those columns are left blank. For example, when the subject did not give a response to a certain inducer, we leave the 15 columns open. It is still possible to determine by computer what inducers the different subjects did not respond to. It is obvious by now that the number of columns needed for each goal object depends on the type of information one wants to analyze. For example, in a study that is solely interested in time perspective and when the researcher is not interested in which answers were given to which inducers, only six columns are needed for each motivational object.

2. The columns needed for one particular goal object form one block and are never dispersed over two cards. If the number of columns left on one card is insufficient to inscribe the whole block of the following goal object, these columns are left open and transcription goes on on the following card. Thus, the number of cards needed for each subject will depend on the amount of personality data inscribed on the first card and on the number of columns needed for each of the following goal objects. If enough columns are left on the first card after inscribing all the personality data, one or more blocks of MIM

data can be transcribed on that card; if not, the columns are left open. On the following cards, there are 75 columns left for MIM data (after 5 columns for identification-codes). When a set of 15 columns is needed for each goal object, the information for 5 objects can be stored on one card. In that case all the columns are used. When only 13 columns are needed for each response, 5 objects can also be stored on one card. But now, the ten last columns of each card are not used.

The number of cards needed for one subject will also depend, of course, on the number of MIM inducers used (only the positive, or also the negative inducers; the complete list, or one of the two shortened versions).

NOTE: *More than one goal object in one sentence completion*

A subject may express more than one motivational object in a sentence completion (cf. p.104). When not more than one goal object for each inducer is coded, the method explained above appplies. When more than one goal object is coded, the following rule can be applied. The numerical codes of the different objects in one sentence completion follow each other on the punchcards. It does not matter that the number of cards will not be the same for all subjects. When the numbers of inducers are coded, this number is repeated for all goal objects expressed to one inducer. Further analysis of the data may use either all the responses or only the first object expressed to each inducer.

Transcribing motivational content codes and time perspective codes on punchcards strongly facilitates detailed analysis of the motivational and temporal components of MIM responses and of the relationships between the content categories and the time categories. The analytic character of the content and temporal codes - their most important characteristic - can be fully exploited by using the computer.

TABLE 15

Numerical code of the symbols of the MIM time perspective analysis (*)

001	T	030	.E2	062	.A0	093	.A2		
002	D	031	E2	063	A0	094	A2		*References to the past and modalities(6)*
003	W	032	E2.	064	A0.	095	A2.		*(special columns)*
004	W	033	E2	065	A0	096	A2		
005	W.	034	E2-	066	A0-	097	A2-		
006	M	035	E2-	067	A0A1	098	A2-		
007	M	036	E2E3	068	A0A1	099	A2O	100	P 300 ←p
008	M.	037	E2E3	069	A0A1-	100	A2O	102-199 ←P(...)(5) 302-399 ←p(...)(5)
009	M-	038	E2E3-	070	A0A1-	101	A2O-		
010	Y	039	E2E3-	071	A0A	102	A2O-	205 ←P(.O) 405 ←p(.O)
011	Y	040	E2A0	072	A0A	103	A2OX	206 ←P(O) 406 ←p(O)
012	Y.	041	E2A0	073	A0A	104	A2OX	207 ←P(EO) 407 ←p(EO)
013	Y·	042	E2A0-	074	A0A-	105	.O	208 ←P(EOE1) 408 ←p(EOE1)
014	Y·.	043	E2A0-	075	.A1	106	O	209 ←P(E1) 409 ←p(E1)
015	Y-	044	E2A1	076	A1	107	O.	210 ←P(E1E2) 410 ←p(E1E2)
016	E(1)	045	E2A1	077	A1.	108	O.	001 ↔
017	.E	046	E2A1-	078	A1	109	O-	002 →
018	E.	047	E2A1-	079	A1-	110	O-	
019	E	048	E2A	080	A1-	111	OX	
020	E-	049	E2A	081	.A	112	OX	
021	E-	050	E2A-	082	A	113	.X	
022	EA1	051	E2A-	083	A.	114	X	
023	EA1	052	.E3	084	A	115	X.	
024	EA1-	053	E3	085	A-	116	X	
025	EA1-	054	E3.	086	A-	117	L	
026	EA	055	E3	087	A0	118	L	
027	EA	056	E3-	088	A0	119	x	
028	EA-	057	E3-	089	A0-	120	x	
029	EA-	058	E3A1	090	A0-	121	U	
		059	E3A1	091	A0X				
		060	E3A1-	092	A0X				

(*) For the footnotes see next page

FOOTNOTES OF TABLE 15

(1) The life periods E_0 and E_1 (0-12 years) as such do not have a numerical code because the MIM can only be used from about the age of 12. However, they are coded between parentheses for objects situated in the past or for references to the past. See the numerical codes 800-803, 207-210, and 407-410.

(2) After the symbol small *l* (open present), the time period during which the expressed goal object applies most can be coded between parentheses. That period must start with the present life period of the subject but it may extend into following periods (combined symbol). Dependent on the age of the subjects, the period between parentheses may take a numerical code from 016 (subject in E) to 111(OX) in the list of future time categories. The numerical code for *l* is 200. The code for *l* followed by a temporal code between parentheses is the numerical code for that time category plus 200. For example, the Code *l(E)* is transcribed as 216, and *l(OX)* as 311.

(3) See note 2. When the small letter *l* is underlined we add 400 to the numerical code for the time category between parentheses.

(4) Very exceptionally, the motivational object must be situated in the past (Code *P*). In such cases the numerical code for the past is punched in the three columns reserved for time perspective (columns 10 to 12 in a block of 15).

(5) The principle is the same as for lower case *l*. The shortest reference to the past is to the Category *D* (the same day or yesterday). The numerical code for *D* is 002. *P(D)* is then coded 602 (002 plus the code for *P* being 600). References to the E_0 and E_1 life periods get a special code because we do not have a numerical code for these time categories in the list (see note 1).

(6) Three columns (13 to 15 in a block of 15) are reserved for implicit and explicit references to the past in sentence completions with a future related goal object . The future time category is punched in columns 10 to 12 (of a complete block of 15 columns). The reference to the past that the subject expresses together with a future goal object is punched in columns 13 to 15.

TABLE 16a

Numerical codes of the symbols of the MIM content analysis
(first part)

001-099: Cat. S		100-199: Cat. SR		200-299: Cat. R		300-399: Cat. C	
001	S	100	SR	200	R	300	C
002	$S(C)$	101	$SR(C)$	201	$R2$	301	$Cint$
003	$S(Cint)$	102	$SR(Cint)$	202	$R2r$	302	Ceg
004	Sc	103	$SRph$	203	$R3$	303	$Calt$
005	Sph	104	$SRcar$	204	Ra	304	$C(Tr)$
006	$Scar$	105	$SRcar(C)$	205	$R2a$	305	$C(Tx)$
007	$Scar(R)$	106	$SRcar(Cint)$	206	$R3a$	306	$C(R2)$
008	$Scar(R2)$	107	$SRcar(L)$	207	$R2ra$	307	$C(R3)$
009	$Scar(R3)$	108	$SRcar(Tr)$	208	$R(E)$		
010	$Scar(C)$	109	$SRcar(Tx)$	209	$R(P)$		
011	$Scar(Cint)$	110	$SR(Calt)$	210	$R(Tr)$		
012	$Scar(L)$	111	$SR(Ceg)$	211	$R(Tx)$	400-499: Cat. $C2$	
013	$Scar(Tr)$	112	$SRapt$	212	$R(L)$		
014	$Scar(Tx)$	113	$SRapt(E)$	213	$Rn(ph)$	400	$C2(af)$
015	$S(Calt)$	114	$SRapt(L)$	214	$Rn(ps)$	401	$C2(ap)$
016	$S(Ceg)$	115	$SRaut$			402	$C2(su)$
017	$Sapt$	116	$SRaut(ec)$			403	$C2(S)$
018	$Sapt(E)$	117	$SRaut(ph)$			404	$C2(Scar)$
019	$Sapt(R)$	118	$SRaut(ps)$	01-19:		405	$C2(SCalt)$
020	$Sapt(R2)$	119	$SRaut(Tr)$	Code for persons		406	$C2(Saut)$
021	$Sapt(R3)$	120	$SRaut(Tx)$	(special columns)		407	$C2(P)$
022	$Sapt(L)$	121	$SR(P)$			408	$C2(C)$
023	$Spre$	122	$SR(E)$			409	$C2(Cint)$
024	$Spre(ps)$	123	$SR(L)$	01	e	410	$C2(Ceg)$
025	$Spre(ph)$	124	$SR(ec)$	02	f	411	$C2(R)$
026	$Spre(ec)$	125	$SR2$	03	p	412	$C2(R2)$
027	$Spre(Tr)$	126	$SR2r$	04	a	413	$C2(R3)$
028	$Spre(Tx)$	127	$SR3$	05	gr	414	$C2(E)$
029	$Saut$	128	$SR(Tr)$	06	pl	415	$C2(L)$
030	$Saut(ec)$	129	$SR(Tx)$	07	in	416	$C2(Tr)$
031	$Saut(ph)$	130	$SR2(Tr)$	08	g	417	$C2(Tx)$
032	$Saut(ps)$	131	$SR2(Tx)$	09	so		
033	$Saut(R2)$	132	$SR3(Tr)$	10	x		
034	$Saut(R3)$	133	$SR3(Tx)$	11	si		
035	$Saut(Tr)$			12	is		
036	$Saut(Tx)$			13	y		
				14	ol		
				15	fy		
				16	fol		
				17	sp		
				18	an		
				19	nat		

TABLE 16b

Numerical codes of the symbols of the MIM content analysis
(second part)

500-599: Cat. C_3		600-699: Cat. E		800-899: Cat. L	
500	$C_3(S)$	600	E		
501	$C_3(Spb)$	601	$E(S)$	800	L
502	$C_3(S,C)$	602	$E(Scar)$	801	$L(E)$
503	$C_3(S,Cint)$	603	$E(SCalt)$	802	$L(pb)$
504	$C_3(Sc)$	604	$E(SCeg)$	803	$L(ss)$
505	$C_3(Scar)$	605	$E(Sapt)$	804	$L(gf)$
506	$C_3(SCalt)$	606	$E(Saut)$	805	$L(nR)$
507	$C_3(SCeg)$	607	$E(R2)$	806	$L(C)$
508	$C_3(Sapt)$	608	$E(R3)$	807	$L(P)$
509	$C_3(Spre)$	609	$E(C)$		
510	$C_3(Saut)$	610	$E(Cint)$		
511	$C_3(SR)$	611	$E(Tr)$		
512	$C_3(SR2)$	612	$E(Tx)$		
513	$C_3(SR3)$	613	$E(l)$	900-949: Cat. Tt	
514	$C_3(SRpb)$	614	$E(w)$		
515	$C_3(SR,C)$			900	Tt
516	$C_3(SR,Cint)$			901	$Tt+$
517	$C_3(SRcar)$			902	$Tt-$
518	$C_3(SRCalt)$				
519	$C_3(SRCeg)$	700-749: Cat. T			
520	$C_3(SRapt)$				
521	$C_3(SRaut)$	700	Tr		
522	$C_3(R)$	701	Tx	950-999: Cat. U	
523	$C_3(R2)$				
524	$C_3(R3)$			950	U
525	$C_3(Rn,ps)$			951	$U?$
526	$C_3(Rn,pb)$			952	Ui
527	$C_3(C)$	750-799: Cat. P			
528	$C_3(Cint)$				
529	$C_3(Ceg)$	750	P		
530	$C_3(Calt)$	751	$P(Spb)$		
531	$C_3(af)$	752	$P(R)$	Code for modalities	
532	$C_3(ap)$	753	$P(R2)$	(special columns)	
533	$C_3(su)$	754	$P(R3)$		
534	$C_3(P)$	755	$P(L)$	01	m
535	$C_3(L)$	756	$P(Lss)$	02	s
536	$C_3(Tr)$	757	$P(E)$	03	d
537	$C_3(Tx)$	758	$P(Ew)$	04	du
538	$C_3(E)$	759	$P(an)$	05	b
539	C_3,H	760	$P(gf)$	06	↔
				07	gf
				08	→
				09	n (before the symbol)
				10	$n+$ (before the symbol)
				11	$n+$ (after the symbol)
				12	np
				13	$/R$

REFERENCES

Aall, A. Die Bedeutung der Zeitperspektive bei der Einprägung für die Dauer der Gedächtnisbilder. In F. Schumann (Ed.), *Bericht über den V. Kongress für experimentelle Psychologie*. Leipzig: Barth, 1912, pp.237-241.

Aall, A. Ein neues Gedächtnisgesetz? Experimentele Untersuchung über die Bedeutung der Reproduktionsperspektive. *Zeitschrift für Psychologie*, 1913, *66*, 1-50.

Allport, G.W. *Pattern and growth in personality*. New York: Holt, 1961.

Argyris, C., & Schön, D.A. *Theory in practice. Increasing professional effectiveness*. San Francisco: Jossey-Bass, 1974.

Argyris, C., & Schön, D.A. *Organizational learning: A theory of action perspective*. Reading, Mass.: Addison-Wesley, 1978.

Barker, R.G. *The stream of behavior*. New York: Appleton-Century-Crofts, 1963.

Battle, E., & Rotter, J.B. Children's feelings of personal control as related to social class and ethnic group. *Journal of Personality*, 1963, *31*, 482-490.

Blatt, S.J., & Quinlan, P. Punctual and procrastinating students: A study of temporal parameters. *Journal of Consulting Psychology*, 1967, *31*, 169-174.

Bochner, S., & David, K.H. Delay of gratification, age and intelligence in an aboriginal culture. *International Journal of Psychology*, 1968, *3*, 167-174.

Bouffard, L. *Contenu et extension temporelle des projets d'avenir en relation avec la frustration (Content and temporal extension of future projects as a function of frustration)*. Unpublished doctoral dissertation, University of Leuven, 1980.

Brim, O. G., & Forer, R. A note on the relation of values and social structure to the life planning. *Sociometry*, 1956, *19*, 54-60.

Cantril, H. *The pattern of human concerns*. New Brunswick, N.J.:

Rutgers University Press, 1965.
Cohen, Jacob. A coefficient of agreement for nominal scales. *Educational and Psychological Measurement*, 1960, *20*, 37-46.
Cohen, John. *Psychological time in health and disease.* Springfield, Ill.: Charles C. Thomas, 1967.
Combs, A.W. A method for analysis for the thematic apperception test and autobiography. *Journal of Clinical Psychology*, 1946, *2*, 167-174. (a)
Combs, A.W. A comparative study of motivations as revealed in thematic apperception stories and autobiography. *Journal of Clinical Psychology*, 1946, *3*, 65-75. (b)
Cossey, H. *Vergelijking van twee methodes in het onderzoek van de menselijke motivatie (A comparison of two methods to study human motivation).* Unpublished master's thesis, University of Leuven, 1967.
Cossey, H. *De intensiteit van menselijke motivaties. Een theoretische en empirische bijdrage (The intensity of human motivations. A theoretical and empirical study).* Unpublished doctoral dissertation, University of Leuven, 1974.
Cossey, J.-M. *Het toekomstperspectief bij jeugddelinquenten. Literatuuroverzicht en vergelijkend onderzoek (The future time perspective of young delinquents).* Unpublished master's thesis, University of Leuven, 1975.
Cottle, T.J. *Perceiving time: A psychological investigation with men and women.* New York: Wiley-Interscience, 1976.
Cottle, T.J., & Klineberg, S. L. *The present of things future. Explorations of time in human experience.* New York: Free Press, 1974.
Craeynest, P. *Constantie of variabiliteit in de inhoud van zinsaanvullingen (Constancy or variability in the content of MIM-sentence completions).* Unpublished master's thesis, University of Leuven, 1967.
De Volder, M. Attitude of male and female college students towards the personal past, present, and future, as a function of experience of success or failure. *University of Leuven Psychological Reports, Research Center for Motivation and Time Perspective*, Nr. 8, 1978.
De Volder, M. Time orientation: A review. *Psychologica Belgica*, 1979, *19*, 61-79.
De Volder, M., & Lens, W. Academic achievement and future time

perspective as a cognitive motivational concept. *Journal of Personality and Social Psychology*, 1982, *42*, 566-571.

Dollard, J., & Auld, F., Jr.   *Scoring human motives: A Manual.* New Haven: Yale University Press, 1959.

Duces, B.   *Bijdrage tot de studie van de intensiteit van motivatie (The intensity of motivations).* Unpublished master's thesis, University of Leuven, 1968.

Eccles, J.C.   *Facing reality. Philosophical adventures by a brain scientist.* New York: Springer, 1970.

Edgerton, S.Y.   *The Renaissance rediscovery of linear perspective.* New York: Harper & Row, 1975.

Fraisse, P.   *Psychologie du temps.* Paris: Presses universitaires de France, 1957 (2nd edition 1967); English translation: *The psychology of time.* New York: Harper & Row, 1963.

Fraisse, P.   Cognition of time in human activity. In G. d'Ydewalle & W. Lens (Eds.), *Cognition in human motivation and learning.* Leuven: Leuven University Press - Hillsdale, N.J.: Erlbaum, 1981, pp. 233-259.

Frank, L.K.   Time perspectives. *Journal of Social Philosophy*, 1939, *4*, 293-312.

Geirnaert, W.   *Dominante tijdgerichtheid: literatuurstudie en empirisch onderzoek (Dominant time orientation. A theoretical and empirical study).* Unpublished master's thesis, University of Leuven, 1976.

Gjesme, T.   Goal distance in time and its effects on the relations between achievement motives and performance. *Journal of Research in Personality*, 1974, *8*, 161-171.

Gjesme, T.   Future-time gradients for performance in test anxious individuals. *Perceptual and Motor Skills*, 1976, *42*, 235-242.

Goethals, J.-M.   *Bewuste dynamische inhouden van gedetineerden: een vergelijkend onderzoek (The conscious dynamic contents of prisoners: A differential research).* Unpublished doctoral dissertation, University of Leuven, 1967.

Goldrich, J. M.   A study in time orientation: The relation between memory for past experiences and orientation to the future. *Journal of Personality and Social Psychology*, 1967, *6*, 216-221.

Heckhausen, H.   Achievement motivation and its constructs: A cognitive model. *Motivation and Emotion*, 1977, *1*, 283-329.

Heimberg, L.K.   *The measurement of future time perspective.* Unpublished doctoral dissertation, Nashville, Tenn., Vanderbilt

University, 1963.

Hoornaert, J. Time perspective: Theoretical and methodological considerations. *Psychologica Belgica*, 1973, *13*, 265-294.

Kastenbaum, R. Cognitive and personal futurity in later life. *Journal of Individual Psychology*, 1963, *19*, 216-222.

Kirk, R.E. *Experimental design: Procedures for the behavioral sciences.* Belmont, Cal.: Brooks-Cole Publishing Company, 1968.

Lefebre, D. *Inhoudsanalyse van de zinsaanvullingen op positieve en negatieve motivationele inductoren bij gedetineerden voor diefstal en moord (The motivational content of positive and negative MIM-sentence completions of prisoners for murder or robbery).* Unpublished master's thesis, University of Leuven, 1969.

Lens, W. *Vergelijkende studie van projectief en direct verbaal materiaal in motivatie-onderzoek (A comparative study of motivational contents in projective and in direct, first person thought samples).* Unpublished doctoral dissertation, University of Leuven, 1971.

Lens, W. *De affectieve attituden tegenover het persoonlijk verleden, heden en toekomst (The affective attitudes towards the personal past, present, and future).* Unpublished manuscript, Leuven, 1972.

Lens, W. A comparative study of motivational contents in projective and in direct, first person thought samples. *Psychologica Belgica*, 1974, *14*, 31-54.

Lens, W. Sex differences in attitude towards personal past, present and future. *Psychologica Belgica*, 1975, *15*, 29-33.

Lens, W., & Gailly, A. Content and future time perspective of motivational goals in different age groups. *University of Leuven Psychological Reports, Research Center for Motivation and Time Perspective*, Nr 10, 1978.

Lens, W., & Gailly, A. Extension of future time perspective in motivational goals of different age groups. *International Journal of Behavioral Development*, 1980, *3*, 1-17

Leshan, L.L. Time orientation and social class. *Journal of Abnormal and Social Psychology*, 1952, *47*, 589-592.

Lessing, E.E. Extension of personal future time perspective, age, and life satisfaction of children and adolescents. *Developmental Psychology*, 1972, *6*, 457-468.

Lewin, K. Sachlichkeit und Zwang in der Erziehung zur Realität. *Die Neue Erziehung*, 1931, *Nr 2*, 99-103.
Lewin, K. *A dynamic theory of personality. Selected papers.* New York-London: McGraw-Hill, 1935.
Lewin, K. Time perspective and morale. In **G. Watson** (Ed.), *Civilian morale.* Boston: Houghton Mifflin, 1942.
Lewin, K. Behavior and development as a function of the total situation. In **L. Carmichael** (Ed.), *Manual of child psychology.* New York-London: Wiley-Chapman & Hall, 1946, pp. 791-844 (Chapter 10).
Lewin, K. *Field theory in social science. Selected theoretical papers* (edited by **D. Cartwright**) New York: Harper & Row, 1951; London: Tavistock Publications, 1952.
Leyssen, M. *Affectieve attitudes tegenover verleden, heden en toekomst in relatie tot motivatie en tijdsperspectief (Affective attitudes towards the past, the present, and the future as a function of motivation and time perspective).* Unpublished master's thesis, University of Leuven, 1974.
Lipman, R. *Some relationships between manifest anxiety, defensiveness and future time perspective.* Stencil, Conference held at the "Convention of Eastern Psychological Association", 1957.
Lotze, R.H. *Grundzüge der Psychologie. Diktate aus den Vorlesungen.* Leipzig: Hirzel, 1881. English translation by **G.T. Ladd**, *Outlines of psychology.* Boston: Ginn, 1886. Chapter on the theory of local signs reprinted in: **B. Rand**, *The classical psychologists.* Boston: Houghton Mifflin, 1912.
Malrieu, Ph. *Les origines de la conscience du temps.* Paris: Presses universitaires de France, 1953.
McClelland, D.C., Atkinson, J. W., Clark, R. A., & Lowell, E. L. *The achievement motive.* New York: Appleton, 1953.
Melges, F.T., & Bowlby, J. Types of hopelessness in psychopathological process. *Archives of General Psychiatry*, 1969, *20*, 690-699.
Menahem, R. L'espace sémantique temporel à différents âges de la vie et sa structuration lors d'une crise suicidaire. *L'Année Psychologique*, 1972, *72*, 353-377.
Meyer-Billet, G. *Bijdrage tot de studie van de objecten der bewuste menselijke motivaties (A contribution to the study of the objects of conscious human motivations).* Unpublished master's thesis, University of Leuven, 1966.

Meyer, G., & Grommen, R. Etude comparative de la motivation et de la perspective temporelle belge chez des groupes tanzaniens et belges. *University of Leuven Psychological Reports, Research Center for Motivation and Time Perspective*, Nr. 6, 1975.

Mönks, F. Zeitperspektive als psychologische Variabele. *Archiv für die Gesamte Psychologie*, 1967, *119*, 131-161.

Moors, S. *Inhoudelijke studie van de verdere motiveringen van bewuste aspiraties. Een onderzoek bij priesterstudenten (The more fundamental goal objects of conscious aspirations).* Unpublished doctoral dissertation, University of Leuven, 1972.

Murray, H.A., Drive, time, strategy, measurement, and our way of life. In **G. Lindzey** (Ed.), *Assessment of human motives.* New York: Rinehart & Company, 1958, pp. 183-196.

Murray, H.A. et al. *Explorations in personality.* New York: Oxford University Press, 1938.

Murray, H.A., MacKinnon, D.W., et al. *Assessment of men.* New York: Rinehart, 1948.

Murthy, H.N. *A study of human motivation in the Indian context.* Leuven: Institute of Psychology, 1963.

Nisbett, R.E., & Wilson, T.D. Telling more than we can know: Verbal reports on mental processes. *Psychological Review*, 1977, *84*, 231-259.

Noterdaeme, Th. *Het tijdsperspectief in de aspiraties: uitwerking van een methode (Time perspective in aspirations: The development of a research instrument).* Unpublished doctoral dissertation, University of Leuven, 1965.

Nowakowska, M. *Language of motivation and language of actions.* The Hague-Paris: Mouton, 1973.

Nuttin, J. *De wet van het effect en de rol van de taak in het leerproces (The law of effect and the role of the task in learning).* Unpublished doctoral dissertation, University of Leuven, 1941.

Nuttin, J. *Tâche, réussite et échec. Théorie de la conduite humaine* (Studia Psychologica). Louvain: Publications universitaires de Louvain, 1953 (3rd ed. 1971).

Nuttin, J. *The future time perspective in human motivation and learning.* Proceedings of the 17th International Congress of Psychology, Washington 1963. Amsterdam: North-Holland Publishing Company, & *Acta Psychologica*, 1964, *23*, 60-82.

Nuttin, J. (with the coll. of **A. Greenwald**). *Reward and punishment* *in human learning. Elements of a behavior theory.* New York-London: Academic Press, 1968.

Nuttin, J. *Time Attitude Scale (T.A.S.).* Unpublished Manual. Leuven Research Center for Motivation and Time Perspective, 1972.

Nuttin, J. *Motivation et perspectives d'avenir.* Leuven: Leuven University Press, 1980.

Nuttin, J. *Motivation, planning, and action. A relational theory of behavior dynamics.* Hillsdale, N.J.: Erlbaum, 1984.

Nuttin, J., & Grommen, R. Zukunftsperspektive bei Erwachsenen und älteren Menschen aus drei sozioökonomischen Gruppen. In **U.M. Lehr & F.E. Weinert** (Eds.), *Entwicklung und Persönlichkeit: Beiträge zur Psychologie intra- und interindividueller Unterschiede.* Stuttgart: Kohlhammer, 1975, pp. 183-197.

Nuttin, J., et al. La perspective temporelle dans le comportement humain. Etude théorique et revue de recherches. In **P. Fraisse et al.** *Du temps biologique au temps psychologique.* Paris: Presses universitaires de France, 1979, pp. 307-363.

Popper, K.R. *Objective knowledge. An evolutionary approach.* Oxford: Clarendon Press, 1975 (first edition 1972).

Pringle, M.K. *The needs of children.* New York: Schocken Books, 1974.

Raynor, J.O. Future orientation and motivation of immediate activity: An elaboration of the theory of achievement motivation. *Psychological Review*, 1969, *76*, 606-610.

Raynor, J.O. Future orientation in the study of achievement motivation. In **J.W. Atkinson & J.O. Raynor,** *Motivation and achievement.* Washington, D.C.: Winston & Sons, 1974, pp. 121-154.

Ricks, D., Umbarger, C., & Mack, R. A measure of increased perspective in successfully treated adolescent delinquent boys. *Journal of Abnormal and Social Psychology*, 1964, *69*, 685-689.

Rizzo, A.E. The time moratorium. *Adolescence*, 1967-1968, *2*, 469-480.

Rotter, J.B. Generalized expectancies for internal versus external control of reinforcement. *Psychological Monographs*, 1966, *80*, whole Nr 1.

Ryan, T.A. *Intentional behavior. An approach to human motivation.* New York: The Ronald Press Company, 1970.

Ryan, T.A. Intentions and kinds of learning. In **G. d'Ydewalle & W. Lens** (Eds.), *Cognition in human motivation and learning.*

Leuven: Leuven University Press - Hillsdale, N.J.: Erlbaum, 1981, pp. 59-85.

Santostefano, S. Assessment of motives in children. *Psychological Reports*, 1970, *26*, 639-649.

Shapira, Z. Expectancy determinants of intrinsically motivated behavior. *Journal of Personality and Social Psychology*, 1976, *34*, 1235-1244.

Shostrom, E.L. *Personal Orientation Inventory*. San Diego, California: EdTS, 1963.

Shostrom, E.L. Time as an integrating factor. In **Ch. Bühler & F. Massarik** (Eds.), *The course of human life. A study of goals in the humanistic perspective*. New York: Springer, 1968, pp. 351-359.

Siegel, S. *Nonparametric statistics for the behavioral sciences*. New York: McGraw-Hill, 1956.

Smith, E.R., & Miller, F.D. Theoretical note. Limits on perception of cognitive processes: A reply to Nisbett and Wilson. *Psychological Review*, 1978, *85*, 355-362.

Stein, M.I. The use of a sentence completion test for the diagnosis of personality. *Journal of Clinical Psychology*, 1947, *3*, 47-56.

Tolman, E.C. *Purposive behavior in animals and men*. New York-London: The Century Co, 1932.

Trommsdorff, G., Lamm, H., & Schmidt, R.W. A longitudinal study of adolescents' future orientation (time perspective). *Journal of Youth and Adolescence*, 1979, *8*, 131-147.

Tulving, E. Episodic and semantic memory. In E. Tulving & W. Donaldson (Eds.), *Organization of memory*. New York: Academic Press, 1972.

Van Calster, K. *Angst en toekomstperspectief (Anxiety and future time perspective)*. Unpublished master's thesis, University of Leuven, 1971.

Van Calster, K. *Analyse, operationalisatie en empirisch onderzoek van de affectieve attitude tegenover de persoonlijke toekomst, het persoonlijk heden en verleden (The affective attitude towards the personal past, present and future: A theoretical analysis and an empirical study)*. Unpublished doctoral dissertation, University of Leuven, 1979.

Vanden Auweele, Y. *Toekomstanticipaties van 15-jarigen in de "Deutsche Demokratische Republik" en de "Deutsche Bundesrepublik"; een interkultureel "kritisch psychologisch" onder-*

zoek *(Future anticipations of 15-year-old subjects in the Federal Republic of Germany and in the German Democratic Republic).* Unpublished doctoral dissertation, University of Leuven, 1973.

Verstraeten, D. *De realiteitsgraad van toekomst-aspiraties. Literatuurstudie en empirisch onderzoek bij een groep adolescenten (Level of realism in adolescent future time perspective. A theoretical and empirical study).* Unpublished doctoral dissertation, University of Leuven, 1974.

Wallace, M., & Rabin, A.I. Temporal experience. *Psychological Bulletin*, 1960, *57*, 213-236.

Weiner, B. *Theories of motivation. From mechanism to cognition.* Chicago: Markham, 1972.

Weiner, B. (Ed.) *Achievement motivation and attribution theory.* Morristown, N.J.: General Learning Press, 1974.

Weiner, B. *Human motivation.* New York: Holt, Rinehart and Winston, 1980.

Wohlford, P. *Determinants of extension of personal time.* Unpublished doctoral dissertation, Duke University, 1964.

Wundt, W. *Grundzüge der physiologischen Psychologie* (Vol. 2) (sixth revised edition). Leipzig: Verlag W. Engelmann, 1910.

PUBLICATIONS
BY MEMBERS OF THE RESEARCH CENTER FOR MOTIVATION AND TIME PERSPECTIVE

Murthy, H.N. *A study of human motivation in the Indian context.* Nanjangud (India): Raman Power Press, & Louvain: Institut de Psychologie, 1963, 166 p.

Daunais, J.P. Approche à l'étude de la motivation humaine. *Revue de l'Université de Sherbrooke*, 1968, *IV*, 95-263.

Noterdaeme, Thérèse. Het tijdsperspectief in de aspiraties. Uitwerking van een methode. *Psychologica Belgica*, 1968, *8*, 15-49.

Lens, W. Fantasie en gedraging in verband met motivatie en projectieve technieken. *Tijdschrift voor P.M.S. Werk*, 1970, *16*, 147-156.

Vinckier-Hellings, Hilde. De genese van de processen van doelstelling en projectvorming. *Psychologica Belgica*, 1970, *10*, 1-24.

Lens, W. Bewuste motivaties van een groep neurotische en een groep normale militairen. *Psychologica Belgica*, 1971, *11*, 45-58.

Gillies, J. & Bauer, R. Cognitive style and perception of success and failure. *Perceptual and Motor Skills*, 1971, *33*, 839-842.

Bauer, R. & Gillies, J. Cognitive style and influence of successes and failures on future time perspective. *Perceptual and Motor Skills*, 1972, *34*, 79-82.

Bauer, R. & Gillies, J. Measure of affective dimensions of future time perspective. *Perceptual and Motor Skills*, 1972, *34*, 181-182.

Grommen, Ria. De tijdsbeleving in de ouderdom. *Leuvens Bulletin L.A.P.P.*, 1972, *21*, 15-23.

Maluf, Maria Regina. A perspectiva temporal em funcao do sexo, da idade, e do nivel socio-economico. Time perspective in function of sex, age, and socio-economic level. A differential study on Brazilian subjects. *Revista de Psicologia Normal e Patologica*, 1972, *XVIII*, nos 3-4, 3-47.

Sopena Alcorlo, A. Motivacion y perspectiva temporal en sus relaciones con la personalidad. *Revista de Psicologia General y Aplicada*, 1972, 955-981.

Verstraeten, Daniëlle. Het realisme van aspiraties. Ontwikkeling van een definitie. *Psychologica Belgica*, 1973, *13*, 89-109.

de Mahieu, W. Le temps dans la culture Komo. *Africa*, 1973, *43*, 2-17.

d'Ydewalle, G. Time perspective and learning in open and closed tasks. *Psychologica Belgica*, 1973, *13*, 139-147.

Hoornaert, J. Time perspective. Theoretical and methodological considerations. *Psychologica Belgica*, 1973, *13*, 265-294.

Van Den Auweele, Y. Toekomstanticipaties van 15-jarigen in de DDR en de BRD: een intercultureel, kritisch-psychologisch onderzoek. *Tijdschrift voor Opvoedkunde*, 1973-1974, *19*, 362-385.

Lens, W. & Atkinson, J.W. Academic achievement in High School related to 'intelligence' and motivation as measured in sixth and ninth grade. *Proceedings of the 18th International Congress of Applied Psychology, Montreal, 1974.*

Lens, W. A comparative study of motivational contents in projective and in direct, first person thought samples. *Psychologica Belgica*, 1974, *14*, 31-54.

Lens, W. Motivatie en intellectuele prestaties. *Tijdschrift voor Opvoedkunde*, 1974-1975, *20*, 19-29.

Lee, Mercedes. Attitudes towards past, present and future among Chinese subjects. *Psychologica Belgica*, 1974, *14*, 297-312.

Lens, W. Sex differences in attitude towards personal past, present and future. *Psychologica Belgica*, 1975, *15*, 29-33.

Breesch-Grommen, Ria. Het tijdsperspectief in volwassenheid en ouderdom: theoretische en empirische bijdragen. *Nederlands Tijdschrift voor Gerontologie*, 1975, *6*, 90-105.

Nuttin, J.(R.) & Grommen, Ria. Zukunftsperspektive bei Erwachsenen und älteren Menschen aus drei sozio-ökonomischen Gruppen. In **Ursula Lehr & F.E. Weinert** (Eds.), *Entwicklung und Persönlichkeit: Beiträge zur Psychologie intra- und interindividueller Unterschiede.* Stuttgart: Verlag W. Kohlhammer, 1975, p. 183-197.

Nuttin, J. (with the collaboration of **W. Lens**). La motivation. In **P. Fraisse, J. Piaget** (Eds.), *Traité de psychologie expérimentale*, vol. V. Motivation, émotion et personnalité. Paris: Presses universitaires de France, 1975 (Third revised edition), p. 5-96.

Atkinson, J.W., Lens, W. & O'Malley, P.M. Motivation and ability: Interactive psychological determinants of intellective performance, educational achievement, and each other. In **W.H. Sewell, R.M. Hauser, & D.L. Featherman** (Eds.), *Schooling and achievement in American society.* New York: Academic Press, 1976, p. 29-60.

Bouwen, R. Anticipation and realization: Attitudes and buying plans in the future time orientation of consumer behor.*Psychologica Belgica*, 1977, *17*, 113-134.

Van Calster, K., De Volder, M. Het verband tussen studiegerichtheid en schooluitslag. *Tijdschrift voor Psycho-Medisch-Sociaal Werk*, 1978, *24*, 49-65.

De Volder, M. Time orientation: A review. *Psychologica Belgica*, 1979, *19*, 61-79.

Nuttin, J. (with the collaboration of **W. Lens, K. Van Calster, & M.**

De Volder). La perspective temporelle dans le comportement humain: étude théorique et revue de recherches. In **P. Fraisse** et al. *Du temps biologique au temps psychologique*. Paris: Presses universitaires de France, 1979, p. 307-363.

Leroux, J. Une mesure de l'extension de la perspective temporelle future. In **P. Fraisse** et al. *Du temps biologique au temps psychologique*. Paris: Presses universitaires de France, 1979, p. 365-377.

Lens, W. Vaardigheden en motivationele variabelen als determinanten van prestaties: een theoretische analyse. In E. De Bruyn (Ed.), *Ontwikkelingen in het onderzoek naar prestatiemotivatie: theorie, meetmethode en toepassing in het onderwijs*. Lisse: Swets & Zeitlinger, 1979, p. 29-56.

Nuttin, J. *Théorie de la motivation humaine: du besoin au projet d'action*. Paris: Presses universitaires de France, 1980, 304 p. (a)

Nuttin, J. *Motivation et perspectives d'avenir*. Leuven: Presses universitaires de Louvain, 1980, 288 p. (b)

Nuttin, J. & Lens, W. Quelques recherches sur la perspective future. In **J.Nuttin**, 1980, p. 97-120.

Nuttin, J. & Lens, W. Manuel du code temporel. *Ibid.* p. 135-186.

Nuttin, J. & Lens, W. Manuel de l'analyse de contenu de la MIM. *Ibid.* p., 187-229.

Gailly, A. & Lens, W. Traitement des informations de la MIM par ordinateur. *Ibid.*, p. 259-264.

Lens, W. & De Volder, M. Achievement motivation and intelligence test scores: A test of the Yerkes-Dodson hypothesis. *Psychologica Belgica*, 1980, *20*, 49-59.

Verstraeten, D. Level of realism in adolescent future time perspective. *Human Development*, 1980, *23*, 177-191.

Lens, W. & Gailly, A. Extension of future time perspective in motivational goals of different age groups. *International Journal* of Behavior Development, 1980, 3, 1-17.

Lens, W. Test anxiety and intellective test performance. *4th International Symposium on Educational Testing, Antwerp, 1980, 24-27 June*. Abstract, p. 85--86.

De Volder, M. & Lens, W. Motivation and future time perspective as possible determinants of study behavior and achievement. *22nd International Congress of Psychology, Leipzig, 6-12 July, 1980*. Abstract Guide, p. 388.

Atkinson, J.W. & Lens, W. Fähigkeit und Motivation als Determinanten momentaner und kumulativer Leistung. In H. Heckhausen (Ed.), *Fähigkeit und Motivation in erwartungswidriger Schulleistung*. Göttingen: Verlag für Psychologie Dr. C.Hogrefe, 1980, p. 129-192.

Lens, W. Professor J.(R.) Nuttin, emeritus: Oprichter van de psychologie te Leuven. *Alumni Leuven*, 1980, *11*, nr. 3, 26-27.

Gailly, A. , Hermans, Ph. & Leman, J. Mediterrane dorpsculturen: het sociaal-cultureel verleden van gastarbeiders in België en Nederland. 1. structuren, symbolen en instituties. *Kultuurleven*, 1980, *47*, 820-840. 2. De socialisatie binnen het mediterrane dorp en haar gevolgen voor de immigratie. *Ibid.*, 928-935.

Lens, W. & Van Calster, K. Toekomstperspectief: een cognitief-dynamische gedragsvariabele. In *Liber amicorum Prof. J.(R.) Nuttin: gedrag, dynamische relatie en betekeniswereld*. Leuven: Universitaire Pers Leuven, 1980, p. 67-87.

Luyten, H. & Lens, W. The effect of earlier experience and reward contingencies on intrinsic motivation. *Motivation and Emotion*, 1981, *5*, 25-36.

d'Ydewalle, G. & Lens, W. (Eds.). *Cognition in human motivation and learning*. Leuven: Leuven University Press; Hillsdale, N.J.: Erlbaum, 1981, 290 p.

Moors, S. Fundamentele behoeften als verder doelobject van bewuste aspiraties. *Psychologica Belgica*, 1981, *21*, 165-180.

Nuttin, J. *De menselijke motivatie. Van behoefte tot gedragsproject.* Deventer: Van Loghum Slaterus, 1981.

Gailly, A. Ethnogeneeskunde en psychiatrie bij Turken. *Kultuurleven*, 1982, *49*, 67-80.

De Volder, M. & Lens, W. Academic achievement and future time perspective as a cognitive-motivational concept. *Journal of Personality and Social Psychology*, 1982, *42*, 566-571.

De Volder, M. Multimotivationele benadering van de relatie tussen motivatie en studieprestaties: een literatuuroverzicht. *Tijdschrift voor Psycho-Medisch-Sociaal Werk*, 1982, *28*, 79-84.

Nuttin, J. Le futur de la motivation et la motivation du futur. In P. Fraisse (Ed.), *Psychologie de demain.* Paris: Presses Universitaires de France, 1982, 119-135.

Van der Keilen, M. L.étendue de la perspective temporelle future et l'attitude à l'égard du présent, du passé et de l'avenir chez les adolescents normaux et handicapés sociaux. Influence du succès et de l'échec expérimental. *Psychologica Belgica*, 1982, *22*, 161-183.

Bouffard, L. Scolarité et motivations individuelles: Etude exploratoire. *Revue Québécoise de Psychologie*, 1982, *3*, 27-46.

Bouffard, L., Lens, W. & Nuttin, J. Extension de la perspective temporelle future en relation avec la frustration. *International Journal of Psychology*, 1983, *18*, 429-442.

Bouffard, L. & Lens, W. Frustrations et motivations concrètes. *Revue Québécoise de Psychologie*, 1983, *4*, 3 - 22.

Lens, W. Fear of failure and the level of performance in ability tests. In J. Helmick & S.B. Anderson (Eds.), *On educational testing: Intelligence, performance, standards, test anxiety, and latent traits.* San Francisco: Jossey-Bass Inc., 1983.

Nuttin, J. *Motivation, planning, and action. A relational theory of* *behavior dynamics.* Leuven: Leuven University Press; Hillsdale, N.J.: Erlbaum, 1984.

Nuttin, J. (with the collaboration of **W. Lens**). *Future time perspective and motivation: Theory and research method.* Leuven: Leuven University Press; Hillsdale, N.J.: Erlbaum, 1985.

AUTHOR INDEX

Aall, A., 40, 209
Allport, G.W., 55, 209
Argyris, Ch., 55, 209
Atkinson, J.W., 213, 215
Auld, F. Jr., 65, 211

Barker, R.G., 21, 209
Battle, E., 32, 209
Blatt, S.J., 27, 209
Bochner, S., 32, 209
Bouffard, L., 61, 71, 77, 209
Bowlby, J., 100, 213
Brim, O.G., 27, 209
Bühler, Ch., 216

Cantril, H., 93, 209
Carmichael, L., 213
Cartwright, D., 213
Clark, R.A., 213
Cohen, Jacob A., 61, 77, 210
Cohen, John, 90, 210
Combs, A.W., 65, 210
Cossey, H., 57, 60, 61, 64, 65, 210
Cossey, J.-M., 77, 210
Cottle, T.J., 11, 32, 210
Craeynest, P., 58, 60, 61, 210

David, K.H., 32, 209
De Volder, M., 11, 27, 37, 38, 93, 210, 211
Dollard, J., 65, 211

Donaldson, W., 216
Duces, B., 65, 211
d'Ydewalle, G., 211, 215

Eccles, J.C., 51, 211
Edgerton, S.Y., 16, 211

Forer, R., 27, 209
Fraisse, P., 12, 14, 26, 40, 211, 215
Frank, L.K., 15, 211

Gailly, A., 81, 89, 201, 212
Geirnaert, W., 96, 211
Gjesme, T., 27, 211
Goethals, J.-M., 56, 58, 59, 60, 211
Goldrich, J.M. 93, 211
Greenwald, A., 41, 214
Grommen, R., 58, 60, 77, 81, 214, 215

Heckhausen, H., 36, 211
Heimberg, L.K., 27, 211
Hoornaert, J., 11, 212

Kastenbaum, R., 27, 212
Kirk, R.E., 95, 212
Klineberg, S.L., 32, 210

Ladd, G.T., 213

Lamm, H., 216
Lefebre, D., 56, 58, 60, 212
Lehr, U.M., 215
Lens, W., 25, 27, 37, 38, 45, 58, 60, 81, 89, 92, 93, 97, 101, 103, 133, 135, 201, 210, 211, 212, 215
Leshan, L.L., 15, 27, 212
Lessing, E.E., 27, 212
Lewin, K., 12, 15, 28, 31, 213
Leijssen, M., 96, 213
Lindzey, G., 214
Lipman, R., 100, 213
Lotze, R.H., 17, 213
Lowell, E.L., 213

Mack, R., 32, 215
MacKinnon, D.W., 214
Malrieu, Ph., 19, 213
Maslow, A.H., 30
Massarik, F., 216
McClelland, D.C., 57, 213
Melges, F.T., 100, 213
Menahem, R., 19, 213
Meyer, G., 56, 58, 213, 214
Miller, F.D., 55, 216
Mönks, F., 40, 214
Moors, S., 56, 62, 63, 214
Murray, H.A., 45, 65, 191, 214
Murthy, H.N., 56, 214

Nisbett, R.E., 55, 214
Noterdaeme, Th., 58, 69, 214
Nowakowska, M., 50, 214
Nuttin, J., 11, 15, 17, 18, 27, 33, 34, 36, 40, 41, 44, 49, 50, 51, 60, 70, 77, 81, 91, 101, 103, 133, 135, 178, 214-215

Perls, F.S., 30

Popper, K.R., 51, 215
Pringle, M.K., 65, 215

Quinlan, P., 27, 209

Rabin, A.I., 40, 217
Rand, B., 213
Raynor, J.O., 27, 215
Ricks, D. 32, 215
Rizzo, A.E., 27, 215
Rogers, C., 30
Rotter, J.B., 29, 32, 209, 215
Ryan, T.A., 41, 215

Santostefano, S., 45, 65, 216
Schmidt, R.W., 216
Schön, D.A., 55, 209
Shapira, Z., 35, 216
Shostrom, E.L., 29, 31, 216
Siegel, S., 62, 216
Smith, E.R., 55, 216
Stein, M.I., 45, 65, 216
Stuart, A., 89

Tolman, E.C., 13, 216
Trommsdorff, G., 15, 216
Tulving, E., 13, 216

Umbarger, C., 32, 215

Van Calster, K., 77, 93, 100, 216
Vanden Auweele, Y., 56, 216
Verstraeten, D., 60, 61, 62, 77, 217

Wallace, M., 40, 217
Watson, G., 213

Weiner, B., 29, 217
Weinert, F.E., 215, 221
Wilson, T.D., 55, 214

Wohlford, P., 16, 217
Wundt, W., 17, 217

SUBJECT INDEX

The following abbreviations are currently used in the Index:
F.T.P. = *Future Time Perspective*
T.P. = *Time Perspective*
T.A.S. = *Time Attitude Scale*

Achievement motivation, 145
Action radius
    F.T.P. and, 20
Atemporal objects, 76, 122, 131
Attitude toward the future, 29, 91-100
Attribution (causal -), 29

Behavior
    impact of F.T.P., 32-38

Calendar time
    coding of, 75, 108, 112, 113-114, 124-125
    extension and, 83-84
    scale of, 75, 108
Categories, *see* Motivational categories, Temporal categories
Code, *see* Motivational content analysis, Temporal code
Computer analysis,
    of motivational content codes, 201-207
    of T.P. codes, 201-207
Contemporaneity principle, 12
Content analysis, *see* Motivational content analysis

Expectancy-value theory, 34-36

Extension of F.T.P., 15, 24-25, 79-90
    impact on behavior, 36-38
    index, 81
    mean F.T.P., 81-89
        biological periods, 84-89
        calendar units, 84
        mean extension, 82-89
        median rank, 82
        social-clock periods, 84-89
    measurement, 79-99
    profile, 25, 79-81

Frequency of expression
    and intensity, 64-65
Future time dimension, *see also* F.T.P.
    attitude toward, 91-100
    conditioning and, 13-14
    historical future, 116
    needs and, 14-15
    psychological origin of, 13
Future time perspective, *see also* Extension, Measurement, T.P.
    aspects, 26-28
    behavioral impact, 31-38
        processes involved, 33-36, 39-40
    construction, 15-23
    definition, 11, 16, 23-26

232 Subject index

    earlier experience and, 31-32
    extension, 79-90
    goal objects and, 24-27
    learning and, 31-32
    objective F.T.P., 19-23, 32, 39
    operational definition, 23-26
    origin, 13-15
    realism, 28-31
    temporal code, 73-77
    temporal extension, 36-40
    vs temporal horizon, 26
    theory of F.T.P., 11-41

Horizon (temporal -), 26, 28, 40

Index of F.T.P., 81
Intensity of motivation
    and frequency of expression, 64-65
Inventory
    of motivational categories (INCAM),
    185-190
        instructions, 186-187
        frequency of objects, 190
        list, 187-188
    of motivational objects (INOM),
      190-199
        instructions, 191-192
        list of objects, 194-199
        subjective intensity measures, 193

Learning
    F.T.P. and, 31-32
Life periods, 70-73
    biological, 75, 113
    calendar periods, 113-114
    social, 108-110
Localization, *see* Temporal localization

Manuals
    of motivational content analysis,
    133-166
        table of contents, 133
        examples, 167-175
    of time perspective analysis, 101-123
        table of contents, 101
        examples, 124-129
Means-end structure
    temporal distance and, 37
Measurement of T.P., 23-26, 67-90, *see also* Manuals, M.I.M.
    average time localization, 68-70
    extension of F.T.P., 79-90
    objective time localization, 68-70
    principle, 67-68
    various methods, 24-25
Mental representation, 12-13
    F.T.P. and, 20-22
    virtual, 20-22
M.I.M., 24, 43-66, 177-183
    additional research possibilities, 63-64
    booklets, 45, 177-178
    computer analysis of, 201-207
    construction, 46-48
    frequency and intensity, 64-65
    inducers, 177
        lists, 180-183
        complete form, 180-181
        shorter form A, 181-182
        shorter form B, 182-183
    instructions, 178-179
    instrument, 45-48, 177
    manual, 135-175
    motivational content, 48-54
        activities, 51-53
        analysis, 49-54, 135-175
        categories, 50-54
        code, 53-54
    negative inducers, 47-48, 180-181

Subject index 233

positive inducers, 46-47, 180
principles, 43-44
reliability, 59-61
social desirability and, 55-56
stability, 61-63
structure, 46-48
T.P. analysis, 43-44, 67-89, 103-131
validity, 54-59
Modalities (code for -), 159-162
Motivational categories, 44-45, 50-54, see also Inventory
activity and work, 145-146
exploration, 153-156
leisure and pleasure, 158-159
main categories, 141
possessions, 157-158
self, 139-143
self-realization, 143-145
social contact, 146-153
test related, 165
transcendental, 156-157
unclassifiable, 165-166
Motivational content analysis
categories, 44-45, 50-54, 139-166
main categories, 141
code for people, 151-153
coding technique, 139-166
by computer, 201-207
examples, 167-175
manual, 133-175
modalities, 159-162
negative components, 162-165
numerical transcription, 201-207
principles, 135-137
purpose of, 49-50
reliability of, 60-61
structure of code, 137-139
Motivational Induction Method, see M.I.M.
Motivational objects, see also M.I.M.

Inventory of, 190-199
Motivation, see also Motivational content analysis
analysis, 48-54
categories of, 50-51
content of, 48-54
measurement by M.I.M., 43-66
objects of, 48-54
stability, 61-63

Negative goal objects, 48, 120-122, 162-165
Negative inducers
responses to, 164-165

Objects, see M.I.M., Temporal code
Open-present, 75-76, 79, 110, 114-116, 123
extension of, 88

Past T.P., 12-13, 30, 39-40, 64, 120
attitude toward past, 91-100
coding of, 76, 111, 119-120, 130-131
F.T.P. and, 14
F.T.P. profile and, 79
past-orientation, 11, 67
references to the past, 76, 119-120
Perspective
concept, 16-17, 41
Profile of F.T.P., 79-81

Reliability
of content code, 60
of the M.I.M., 59-63
of the T.A.S., 93, 98-99
of temporal code, 77

Scale

time attitude, 29, 91-100
Signs, see Temporal signs
Social clock, 75, 113-114
    extension of F.T.P. and, 84-89
    temporal codes and, 108-109
Social desirability, 55-56
Stability
    of temporal code, 77
Subjects
    life periods, 70-73
Symbols, see also Temporal code
    meaning, 107
    repetitive acts, 118-119
    specifications, 117-118
    temporal, 75-78, 107-125
        biological, 75
        calendar, 75
        open present, 75-76
        social, 75

Temporal categories, see also Manual, Measurement, Temporal code, Temporal signs, 17-19
    examples, 124-129
    global categories, 123
Temporal code, see also Manuals
    atemporal objects, 76-77, 122-123
    biological units, 75
    calendar units, 75
    categories, 107
        after death, 110-111, 116
        calendar periods, 108, 113-114
        combined periods, 116-117
        life-period, 109-110, 113-114
        open-present, 75-76, 79, 88, 110, 114-116, 123
        past, (see Past T.P.)
        social & biological, 108-109, 113
    coding technique, 111-123
    coding units, 104-105
    context, 104-105
    specification, 117-118
    computer analysis, 201-207
    density of temporal categories, 78
    numerical transcription, 201-207
    objects for others, 122
    reliability, 77
    social units, 75
    stability, 77
    symbols, 75-78
Temporal dimensions of behavior, 12-13
Temporal localization of objects, see also Manuals, Measurement of T.P.
    distortions, 19
    objective localization, 67-70
    origin, 17
    personal localization, 19
    points of reference, 19
Time attitude, 29, 91-100
    definition, 11, 16, 24, 38
    measurement (T.A.S.), 91-100
    revised T.A.S., 97-100
    thematic differences, 93-97
Time competence, 28-31
Time integration, 28-31
Time orientation, 11, 15-16, 24, 67
Time perspective, see also F.T.P., Manual, Measurement
    action radius, 20
    active T.P., 26-28
    behavior and, 22, 32-38
    cognitive T.P., 27-28
    concept, 11, 19-23
    construction, 15-23
    global T.P., 26
    impact on behavior, 22, 32-38
    learning and, 131-132
    measurement of, 23-26, 67-90

types of method, 24-25
objective T.P., 23-25, 32, 39
operational definition, 23-26
presence of objects in, 19-22
    processes involved, 19-22
psychotherapy and, 30-31
    criticism, 30-31
realism of T.P., 28-31

specific T.P., 26
theory, 11-41
vs time attitude, 11
vs time orientation, 11
total T.P., 26

Validity, *see* M.I.M.

Louvain Psychology Series:
Studia Psychologica

(The Louvain Psychology Series "Studia Psychologica" was founded in 1953 by Albert Michotte and Joseph Nuttin. It is the continuation of the "Etudes de Psychologie" founded in 1912 by Albert Michotte).

A. Michotte, *La perception de la causalité*, 1954, VIII+306 p., 2ᵉ éd. 350 F.

A. Chapanis, A. Lucas, E. Jacobson, N.H. Mackworth, L. Ancona & G. Iacono, *L'automation. Aspects psychologiques et sociaux*, 1960, 122 p. 180 F.

M. Leblanc, *La personnalité de la femme katangaise. Contribution à l'étude de son acculturation*, 1960, 406 p. 610 F.

J. Nuttin, *Psychology in Belgium*, 1961, 80 p., 7 photos. 120 F.

G. Thinès, *Contribution à la théorie de la causalité perceptive. Nouvelles recherches sur l'effet-entraînement*, 1961, 124 p. 180 F.

A. Michotte & coll., *Causalité, permanence et réalité phénoménales. Études de psychologie expérimentale*, 1962, 610 p. 620 F.

A. Fauville, *Perception tachistoscopique et communication*, 1963, 56 p. 90 F.

A. McKenna, *Les réactions psycho-physiologiques à la douleur*, 1963, 126 p. 190 F.

J. Nuttin, *De verstandelijke begaafdheid van de jeugd in de verschillende sociale klassen en woonplaatsen* (with English summary: Primary mental abilities in children as related to parental educational and occupational level, rural living conditions, and sex), 1965, 94 p. 140 F.

G. Vandendriessche, *The parapraxis in the Haizmann Case of Sigmund Freud*, 1965, XXXII-192 p. 320 F.

A. Michotte, G. Thinès, G. Crabbé, *Les compléments amodaux des structures perceptives*, 2ᵉ éd., 1967, 56 p. 90 F.

L. Delbeke, *Construction of preference spaces. An investigation into the applicability of multidimensional scaling models*, 1968, VI-182 p. 410 F.

M.J. Vansina, *Het super-ego : Oorsprong en ontwikkeling van S. Freud's opvattingen over het normerende en het morele in de mens*, 1968, 307 p. 450 F.

J. Nuttin, *Psychoanalyse et conception spiritualiste de l'homme*, 5ᵉ éd., 1968, 368 p. 240 F.

J. Nuttin & B. Beuten, *Handleiding bij de persoonlijkheidsinventaris MMPI*, 2ᵉ herwerkte uitgave, 1969, 112 p. 240 F. Met angstschalen. 360 F.

M.Th. Knapen, *L'enfant Mukongo. Orientation de base du système éducatif et développement de la personnalité*, 2ᵉ éd., 1970, 202 p. 300 F.

W. Smet, *Het Rorschach-onderzoek bij 10- à 15-jarige jongens: Handleiding en normen*, 1970, 224 p.   390 F.

J. Nuttin & K. Swinnen, *Overgang naar het middelbaar onderwijs*, 2e druk, 1971, 180 p.   270 F.

J. Nuttin, *Tâche, réussite et échec. Théorie de la conduite humaine*, 3e impression, 1971, 530 p.   480 F.

J. Nuttin, *Psychoanalyse en persoonlijkheid*, 7e herziene uitgave, 1974, 245 p.   325 F.

L. Knops, *Handleiding bij de analytische intelligentietest voor kleuters van 5 tot 6 jaar*, 2e herwerkte uitgave, 1974, 64 p.   150 F.

J. Stinissen, *Handleiding bij de Collectieve Intelligentietest voor elf tot vijftienjarigen*, 2e herz. uitg., 1972, 75 p. 200 F.

G. Smets, *Aesthetic Judgment and Arousal: An Experimental Contribution to Psycho-aesthetics*, 1973, XXII-106 p. 318 F.

J.M. Nuttin Jr., *The Illusion of Attitude Change: Towards a Response Contagion Theory of Persuasion*, 1975, 236 p., hardcover. In co-publication with Academic Press).   580 F.

Carl Rogers & Marian Kinget, *Psychothérapie et Relations Humaines. Théorie et pratique de la thérapie non directive*:
— Vol. I: *Exposé général*, 7e éd., 1976, 333 p.   360 F.
— Vol. II: *La pratique*, 7e éd., 1977, 260 p.   360 F.

J. Nuttin, *Motivation et perspectives d'avenir*, 1980, 292 p. (In co-publication with Lawrence Erlbaum Associates).
   750 F.

A. Vergote & A. Tamayo, *The Parental Figures and the Representation of God. A Psychological and Cross-cultural Study*, 1980, 255 p., hardcover. (In co-publication with Mouton Publishers).   900 F.

J. Nuttin, *Motivation, Planning, and Action. Relational theory of behavior dynamics*, 1984, 251 p., hardcover. (In co-publication with Lawrence Erlbaum Associates).   1200 F.

J. Nuttin, (with the collaboration of W. Lens), *Future Time Perspective and Motivation. Theory and research method*, 1984, 235 p., paper. (In co-publication with Lawrence Erlbaum Associates).   850 F.

Among the psychological volumes published outside the Series and still available are mentioned:

Miscellanea Psychologica Albert Michotte (original contributions by Bartlett, Burt, Fraisse, Hunter, Janet, Köhler, Piaget, Piéron, Terman, Thorndike, etc.